VIETNAM War SPEAK

?

By

WILLIAM W. STILWAGEN

Harbor Site Books

First Edition: 2020

Vietnam War Speak

A **Harbor Site Books** Publication

ISBN: 978-1-7356763-0-2

Library of Congress Control Number: 2020916823

Harbor Site Books
17467 Plantation Road
Onancock, VA 23417

Printed in the United States of America

DEDICATION

This book is dedicated with respect and admiration to the men and women of the United States Armed Forces who forged the words of the Vietnam Era.

They will forever be defined by how they fought, how they suffered, and ultimately, how they prevailed.

May they live eternally in glory.

Their language is shared herein.

CONTENTS

SECTION 1: GLOSSARY

SECTION 2: APPENDIX

FOREWORD

Nothing in the early life of the nearly three million young Americans who served in Vietnam remotely prepared them for the incredible and traumatic effect of serving in the combat zones in a long, frustrating and brutal war. Nor did the training they received once they joined the military do any more than hint at what awaited them. This is, of course, even more the case for the approximately 500,000 men who actually met the enemy face to face. As Karl Marlantes put it in his book, "What It Is Like To Go To War", the military did a good job of teaching its recruits how to kill, but did not even attempt to teach them how to deal with killing another human being. Nor did their training prepare these young men for the hot, humid, insect ridden, water deprived, disease causing environment in which most of them would live for months on end. Nor were they ready to deal with the excruciating experience of observing their comrades shot or blown up a few feet away from them. This all means that these normal, mostly law abiding American young men were in for a dramatically life-changing experience that would change them forever in ways they could not even begin to imagine.

So, it is not surprising that these soldiers, sailors, airmen and Marines communicated in a language that was unique to their circumstances. Whether talking about their tactical situation, or their hopes and dreams, or their fears and discomfort, they used words and phrases that were peculiar to their situation. As we have read books about the Vietnam War, we have been exposed to some of this language. But there was no guide to this jargon before now. That is until Bill Stilwagen compiled this book. It contains no history, nor any interviews with participants, nor an analysis of decisions or strategy or tactics. It is simply a compilation of the words and phrases used on a daily basis by the participants in the War. You will learn a bit about the weapons used and units that fought the War. You will see the multiple words created and used to dehumanize the enemy, as it is much easier to kill a subhuman than to kill someone like you. You will be exposed to the bizarre and often profane humor of these men who were spirited away from their American lives to a totally foreign and radically incomprehensible universe. You will not find "PTSD" explained, for that acronym nor its full phrase even existed then, at least not in the minds of these men. If they referred to the strange behavior that became known as PTSD, it was by the different names used in previous wars. Or maybe they just said that their comrade was "boocoo dinky dow". (Look it up). Other foreign sounding phrases like "Rat tiec ve cau do" (sorry about that) were used as a poor expression of the fatalism or ennui that was felt by these men marooned in a strange world.

Many of these words and phrases are still used by the veterans of that War when they communicate with each other. For they never got over their experience in Vietnam. They are unable to describe their year in hell to their friends and family who did not share their experience, so they do not really try. It is only when communicating with fellow veterans that they are comfortable revisiting that year long, long ago in the language that they used then. Bill Stilwagen was one of them then, and has spent many years conducting tours of veterans back to Vietnam, where they revisit the sites that live in their memories and converse in their peculiar language. Bill has been a guide for more than sixty such trips and been party to hundreds of such conversations, so he is especially qualified to introduce you to this strange language. He has done all of us a favor by compiling this thesaurus. Thank you, Bill.

Tom Esslinger

Tom served in Vietnam as a Marine infantry officer after graduating from Yale University. He was Commanding Officer of two different Marine rifle companies, including one that spent the Tet Offensive on Hill 881S outside the Khe Sanh Combat Base during the 1968 Siege of Khe Sanh. He has revisited Vietnam three times since 1998, all under the expert guidance of Bill Stilwagen.

PREFACE

It is amazing that most of us survived. Not the war, but that we outlasted the abuse wrought upon us by an ungrateful nation.

Our country and its gutless politicians had effectively abandoned its Vietnam warriors. We were discarded as veterans and became outcasts in our own country. There is a saying: *"A veteran is someone who, at one point in his life wrote a blank check made payable to The United States of America, for an amount up to and including his life."* The government cashed in over 58,000 of those checks. But after the war, the uncashed checks were torn up and thrown back in our faces. Vietnam veterans were vilified.

It would take nearly a decade before we were recognized for our service with a national veterans' memorial that we had to fund and build for ourselves. It would take yet another decade before we were accepted back into society as true veterans. Untold tens of thousands would die within that time, due to depression-enhanced diseases, suicide, and hard living.

As I write this, I am grateful that our nation has finally learned to not blame the warriors for the wars. Military members returning from recent conflicts are treated with respect and honor for placing their lives and futures on the line. If the public has learned its lesson from the shameful way Vietnam veterans were treated, so be it. Our sacrifice and suffering, in that regard, has not been in vain.

I hope you will discover some understanding and insight within these pages.

It was written for those of you who care, and for all who come after you.

ACKNOWLEDGEMENTS

No book is ever written entirely on one's own. After many months of mind-boggling work, and once I thought it perfect, I provided drafts of what I presumed was a finished book to those veterans whose knowledge should give me the stamp of approval. Then came the corrections; the additions; the typo fixes; the contextual suggestions; and on and on. Just when I thought I was done, they pulled me back in, determined for me to get it spot-on.

I am grateful for their learned and fastidious amendments and adjustments (despite their obnoxious and boundless mockery) until I got it precise and ready for publishing.

In alphabetical order by last name, these persons are:

Ron (Recon Ron) Huegel, USMC: Ron joined the Marine Corps in June 1968 at age 17. On 3 June 1970, while serving with 1st Force Recon Company, he was seriously wounded in an attack on the Thuong Duc Special Forces Camp. He also served with Bravo Company, 1st Recon Battalion. After Vietnam, Ron served with the Army Security Agency as a Voice Intercept Operator in the German language, where he monitored East German radio traffic. He was medically-retired (in part due to aggravated Vietnam wounds) from the Army in November 1976, after which he built a career in the travel industry. Ron (multi-lingual in English, Vietnamese, and German) stays busy writing a series of books on Vietnam battle sites and locations of Marine Reconnaissance and Special Forces camps. He is a true stickler for detail which is so important in a book like this. Ron also added the Vietnamese spelling for many of the entries.

Frederick Kraemer, USAF, RET: Fred spent 20 years in the United States Air Force, serving in the Vietnam War, Thailand, New York, and California, to name just a few. After retirement, Fred continued to serve America for another 27 years as a High School instructor in the Air Force JROTC program at Osceola High School in Kissimmee, Florida.

Dave Macedonia, 101st Airborne, USA: Dave was drafted into the Army in August of 1969. He was part of the 1971 incursion into Laos, better known as Lam Son 719. Also, in Vietnam, he served as a rifleman and squad leader with Delta Company, 1/501st Infantry, 101st Airborne Division, and as an advisor with Lima Company, 75th Rangers, 101st Airborne

Division attaining the rank of Sergeant E-5. After returning home, he became a Special Education teacher and Administrator with the Arlington, Virginia County Schools, and a coach/swimming instructor. Since his retirement from teaching, he became a Director of Vietnam Battlefield Tours (VBT), a non-profit tour company that leads veterans, their family members, and historians to the jungles and battlefields of Vietnam. He continues today as a tour guide for VBT.

John Strunk, Kilo/3/5, USMC: John was a Grunt. That noble title should be all I need to say, but I shall elaborate. John arrived in Vietnam in the Fall of 1966 at age 18 as a rifleman with Kilo Company, 3rd Battalion, 5th Marine Regiment, of the 1st Marine Division. By the Spring of 1967 he had become a team leader as the regiment moved into Que Son Valley, and John spent the rest of his tour fighting the 2nd NVA Division. Not long after his second wounding, he rotated home to discover a society that had changed and had abandoned his understanding of citizenship and the responsibility of being a free people. By the Grace of God, John was blessed with a loving wife, and he received Christ as his Lord. The young couple had three children and the son became the Commander of the USS Wyoming. John was able to return to the Que Son Valley in 2010 under the capable guidance of Bill Stilwagen. John authored the acclaimed non-fiction, Vietnam War book entitled, "We Walked Across Their Graves."

INTRODUCTION

If there ever was a time worthy of its own vernacular, it is the Vietnam Era. A generation split by peace and war spawned a library of lingo, slang, and fresh new words like no era before it.

The warriors found that the torch had been passed to them by an assassinated president. It motivated them to bear any burden and to pay any price. The best of this generation did ask what they could do for their country. Nine million served in uniform; 2.8 million served in Vietnam. Over 58,000 gave all they had for the home they loved. Over 300,000 spilled their blood in Southeast Asia. No one came home unscathed. Everyone paid a price.

They bore the burden of their generation. And they created a second language to augment what they brought with them.

In this book, you will find over 2,500 terms, acronyms, jargon, slang expressions, and various lingo transported to you by Vietnam servicemen who showed remarkable insight into their own generational place in history. These are the words with which they communicated, especially at the enlisted level. They took the brunt of all the horror the war had to give.

It is imperative for any serious historian, future descendant, or any interested person to grasp the speech variants of the day and thus put them into proper context for complete understanding. Heed the words of Francis George Steiner who declared, *"When a language dies, a way of understanding the world dies with it, a way of looking at the world."* You are likely to find the meaning here. Within these entries comes common lingo, from 1960's slang to the obscure vernacular that was so important in the daily struggles of Soldiers, Sailors, Airmen, and Marines in Vietnam. Included too, are military acronyms, terms, and some downright humorous gems of expressions. Many of the terms and idioms herein have found their way into today's linguistic norms.

Many of these snippets of created language came about as a way to help these warriors excel in battle or came about as a way to help them cope emotionally with the everyday horrors of the war. Some were just made up to entertain each other. Enter their world and learn from them.

"I dream of lost vocabularies that might express some of what we no longer can."

— Jack Gilbert, <u>The Great Fires</u>

1

Perhaps you have read accounts written by veterans, or maybe you've listened to interviews, or seen movies or documentaries depicting the war. Some of the terms used may have been confusing or just plain impossible to understand . . .

- *Recon's in Indian territory, snooping and pooping; they just broke squelch.*
- *Call in the fast-movers and have them lay in some shake `n bake.*
- *Spooky lit `em up last night.*
- *The ma deuce wasted nine dinks in the wire, and we had a WIA/T&T.*
- *Larry went to the Steam & Cream and got an Article 15 and a drip for his trouble.*
- *The LZ was hot, so the log bird gave us a kick out.*
- *Lost my hootch girl to the MOOSE.*

Their words were colored with beauty, sadness, hilarity, horror, ridicule, and irony.

It is not a language for the timid nor the politically correct.

But it is the language of reality.

A hard reality.

"It is not the literal past, the 'facts' of history, that shape us, but the images of the past embodied in language."

— Brian Friel

SOME THOUGHTS ON HOW TO USE THIS GLOSSARY

This book was compiled within the context of America's Vietnam fighting war (1965-1973).

Some terms live on today; some have evolved into new meanings. But these are as they were back then.

For example, back in the day there was no such term as PTSD. Therefore, you will not find it in this book.

Some words were in existence but had fallen out of use in their original meanings. For example, a *Sortie* is an archaic term meaning a short, quick attack by troops from a defensive position. That *Sortie* expression was never used in Vietnam in that context. The *Sortie* used in Vietnam is referred to as an aviation expression only.

Another illustration would be the *TOW* missile. It was first used in Vietnam in 1972 during the Easter Offensive to destroy Soviet tanks. Only the Vietnam version of the *TOW* is used here. Subsequent *TOW* variants (and there are many) are not cited.

If an entry is universally known, only its base component is used and not all its revisions. An example would be the *M16* Standard Infantry Rifle. Its various models, such as *M16E1* or *M16A1,* are not used.

If a term is better known by its nickname than its military nomenclature name, I use the former for the descriptive line. For example, "Claymore" is a more recognized term than "M18A1." Therefore, I put the description at *Claymore* and just *See Claymore* at the *M18A1* listing.

If terms within an entry explanation are capitalized (but not *italicized*), they can usually be cross-referenced elsewhere in the listings.

Acronyms are always fully capitalized.

Non-acronyms are not fully capitalized.

Enjoy . . .

GLOSSARY

A

A-1: Douglas *Skyraider*; single-seat, piston single-engine, attack propeller aircraft; aka Spad

A-4: Douglas *Skyhawk*; single-seat, delta-winged, single-turbojet, subsonic, 670mph, light attack fighter aircraft

A-6: Grumman *Intruder*; twin seat, twin-jet, subsonic all-weather, carrier-capable, attack aircraft

A-7: Vaught *Corsair II*; single-seat, single-turbojet, subsonic, carrier-capable, light attack fighter aircraft

AA Gun: Anti-Aircraft Gun

AA: Anti-aircraft

AAA: Anti-Aircraft Artillery

AAR: See After Action Report

AB: Ambush

ABN: Airborne

Aboard: On a ship or on base; also *with us*

AC: Aircraft Commander

AC-119: Fairchild *Shadow/Stinger* fixed-wing gunship used for close air support and interdiction; aka Flying Boxcar

AC-130: A gunship version of the C-130, nicknamed *Spectre*

ACAV: Armored Cavalry Assault Vehicle; M113 with a modification kit of shields, a turret, and extra machineguns

Acid: See LSD

Across The Fence: Lingo for past the border from SVN into NVN or into Laos or into Cambodia; aka Over The Fence

Actual: Radio call-sign suffix identifying the leader of any size element in the field

Admin: Short for Administration, either an office or a duty

ADSID: Air-Delivered Seismic Intruder-detection Device; shaped like a large, camouflaged lawn dart

AF: Air Force

AFB: Air Force Base

Afro: A tight-curl hairstyle picked out into a round shape popular with African-Americans

Aft: Naval term for rearward

After Action Report: Written retrospective analysis, including a full narrative, strengths/weaknesses, statistics, etc.

Afterdeck: Open area at the stern of a ship; aka Fantail

AFVN Armed Forces Vietnam Network; military-run, popular music/public service announcement, radio station

Agency, The: See CIA

Agent Orange: Powerful herbicide sprayed on vegetation to deny enemy cover; catch-all term for all defoliants

AGL: Above Ground Level; an aircraft's altitude above the ground measured in feet

AGM-45A: See Shrike

Agreement on Ending the War and Restoring Peace in Vietnam: The official title of the Paris Peace Accords

A-Gunner: Assistant gunner

Ah Shit Valley: Contemptuous term for the despised, rugged, enemy-infested A Shau Valley, A Sầu

AH: Attack Helicopter

AH-1: See Cobra

AHB: Assault Helicopter Battalion

AHC: Assault Helicopter Company, Army, consisting of Hueys and Gunships

AID: See USAID

Aids To Navigation: See ATON

AIM-9: See Sidewinder

Ain't No Thang / Ain't No Big Thang / Ain't Nothin' But A Thang: Slang for nothing to worry about, ignore it

Air America: Airline operated by the CIA for photo reconnaissance and various covert activities

Air Cav / Air CAV: Air Cavalry; helicopter-borne infantry and gunship assault teams; see also Airmobile

Air Picket: Any airborne asset that detects, tracks, and reports enemy aircraft and missiles

Air: Fixed wing aircraft fire

Airborne: Soldiers who are parachute jump-qualified; paratrooper

Airburst: A munition detonation in the air so as to disperse shrapnel over a much larger area

Airedale: Any person assigned to an aircraft unit; term could be venerating or sardonic, depending on context

Airmobile: A military unit whose members are transported to combat areas by helicopter

AIT: Advanced Individual Training; focused training in one's MOS received immediately after Army basic training

AK, AK-47: Alexei Kalashnikov (AK) standard Soviet infantry/service rifle

AK: Casualty category term meaning *Above-the-Knee* traumatic amputation

All Hands: Everyone in a unit

Allies: Nations who sent forces to SVN: U.S., South Korea, Thailand, Australia, New Zealand, and Philippines

Alpha Boat: See ASPB

Alpha Bravo: Ambush; aka AB

Alpha Sierra: Radio code for *All Secure*

Alpha: Automatic Ambush; a series of claymore mines that fire simultaneously when tripped

Alphabet, Military: See addendums

AMB: Ambush

Ambulatory: A casualty who can walk on his own

Ambush: An attack suddenly sprung from a hidden position against unsuspecting enemy combatants

American Red Cross: SRAO NGO that, in part, supports and provides comfort to U.S. troops worldwide

American War: Term the Vietnamese use for the *Vietnam War*

AMF: *Adios Motherfucker*

AmGrunt: Lingo for a person assigned infantry duty while a member of an AmTrac unit

Ammo Dump: An explosives storage area where ordnance is kept until distribution; aka Bomb Farm

Ammo: Ammunition

AmTrac: Amphibious Vehicle, Tracked; it can land Marines from water and travel over land; see LVT

AmTracker: A member of an AMTRAC unit

AN/PRS-4: Mine detector used by Combat Engineers to find mines buried less than 20" deep

AN/PVS-2: Image Intensification Device; first-generation night vision scope; see Starlight Scope

AN/TPS-25: Doppler ground-surveillance radar detecting movement direction/#'s/coords of enemy; aka Tipsy 25

Angel Track: An APC used as an ambulance or field aid station

ANGLICO: Air & Naval Gunfire Liaison Company; Marine unit that coordinates air support and Navy guns

ANZAC: Australian and New Zealand Armed Corps

ANZUS: Communist-expansion deterrent agreement between Australia, New Zealand, and the U.S.

Ao Dai: Traditional Vietnamese dress (high-collared silk blouse, pajama-like pants, long two-panel skirt)

AO: Aerial Observer

AO: Area of Operations

AO: Artillery Observer

AOD: Administrative Officer on Duty

AOR: Area Of Responsibility; large geographical zone assigned strategic command accountabilities; see also TAOR

AP: Armor Piercing

Apache Team: SF six-man intelligence-gathering and reconnaissance squad

APC: Armored Personnel Carrier; see M-113

APD: Anti-Personnel Detector; helicopter-mounted odor-sampling device used to discover enemy groupings

Ape / Go Ape / Went Ape: A ranting, emotional display

APM: All Pilot's Meeting

APO: Army Post Office; mail sent to Army/Air Force personnel in Vietnam went via APO, San Francisco; see also FPO

AR-10: The original ArmaLite semi-automatic, battle rifle, forerunner in the development of the AR-15

AR-15: The Colt semi-automatic, battle rifle, forerunner in the development of the M16 assault rifle

ARA: Aerial Rocket Artillery; gunship concept bringing rocket artillery with great accuracy

Arc Light, Arclight: Code name for massive B-52 Carpet Bombing from as high as 50,000 feet

ARM: Anti-Radar Missile; HE missile with radar-homing device to guide it to, and kill, radar units and installations

Armored Personnel Carrier: See M113

Armpit Sauce: See Nuoc Mam

Army Mud: Coffee

ARMY: Army sardonic term meaning *Airforce Rejected Me Yesterday*

Article 15: Non-Judicial Punishment meted out by commanding officers for minor offenses; see addendums

Artillery: Crew served big guns (howitzers, etc.) that launch large-caliber projectiles against distant targets

Arty: Short for Artillery

ARVN: Army of the Republic of (South) Vietnam; sometimes a single SVN soldier; see Marvin The ARVN

As You Were: An order to resume/continue your previous activity

ASAP: As Soon As Possible

ASEAN: Assoc. of Southeast Asian Nations; to hasten social and economic progress, and foster peace and security

ASHC: Assault Support Helicopter Company

Ashore: Most any place that is not on a Naval or Marine Corps installation

ASPB: Assault Support Patrol Boat; 50' steel-hull, heavily-armored, riverine craft; aka Alpha Boat

Asshole To Bellybutton: Line up tight with one's gut close to the anus of the person before them; aka Nut To Butt

Assholes And Elbows: Extremely busy; in a chaotic hurry

Astern: Navy term for toward the back of a ship; to move rearward

ATC: See Tango Boat

A-Team: 10- or 12-man Green Beret units that trained and led CIDG units

ATON: Aids To Navigation; buoys of all kinds, range and wreck markers, lighthouses, etc. on waterways and oceans

Attrition: The sustained assault on an enemy to reduce his effectiveness to where he can no longer wage war

Aussie: Australian soldier; aka Digger

Automatic Weapon: General reference to any gun or rifle that will continue to fire as long as the trigger is engaged

Autorotation: A unstable glide precipitated by air moving over a helicopter's rotors when the engine fails inflight

AVLB: Armored Vehicle Launched Bridge; M60 portable bridge-laying unit; 60' scissors-type bridge

AW: Automatic Weapon

AWOL: Absent Without Official Leave; any absence from a post or duty with intent to return; see also UA

Azimuth: Horizontal line measured clockwise in 1° increments of a 360° circle from North on a compass

B

B40: See RPG-2

B41: See RPG-7

B-5 Front: Communist operational code name for their military actions in Quang Tri/Thua Thien Provinces in SVN

B-52: Boeing *Stratofortress*; 8 turbofan, high-altitude, 525mph, heavy bomber; aka BUF, BUFF, and Dump Truck

Ba Muoi Ba: Vietnamese beer; brewed using formaldehyde for clarity and to extend shelf life, Ba Mươi Ba; aka 33

BA: Base Area

Babysan: Bastardized Japanese word meaning honorable baby; a Vietnamese infant or toddler

Bac Si: Vietnamese word for doctor; a Corpsman or medic

Back Blast: Hi-pressure, hi-temperature, and sometimes deadly, discharge following a rocket launch

Bad Conduct Discharge: See addendums

Bad: Vernacular for good or awesome; sometimes powerful

Badass: Slang for tough guy

Bag: See What's Your Bag

Baht: The basic monetary unit of Thailand currency

Ballgame: Lingo for contact with the enemy

BAM: Broad-Ass Marine; derogatory name for a female Marine; see also HAM

Bamboo Whip: See Malayan Gate

Ban The Bra: A feminist movement slogan implying sexual liberation and barring constraint of women's rights

Banana Clip: Arc-shaped rifle magazine

Banded Krait: Deadly, black & white, triangle-shaped, viper whose venom kills in less than 1-hour; aka Two-Stepper

Bandolier: Strap-on, sash belt with a series of pockets worn across the chest for holding ammo clips or magazines

Bangalore: M1A1, 5', 13lb, explosive tube, linkable, hand-laid or rocket-launched, breach weapon

BAR: M1918A2 Browning Automatic Rifle; 20-round magazine, 600 rounds per-minute, 19lb, light machinegun

Barbeque: Radio code used when requesting a napalm attack

Barracks: A unit's living quarters; building used to house military persons

BAS: Battalion Aid Station; forwardmost treatment facility to a battle area that is medically staffed

Base Camp Commando: A soldier assigned to a safe rear area, not exactly a REMF, but close to it

Base Camp: A central encampment that provides supplies, shelter, fire support, and communications

Base Piece: Heavy gun (artillery or mortar) in the center of a battery; fires the first shot of a fire mission

Baseball Grenade: A U.S. 14 oz, baseball-shaped, fragmentation, hand-thrown explosive device; aka M67

Basic: Recruit training; Boot Camp

Basket Boat: A round, woven-rattan, pitch-plastered, one-paddle powered, Vietnamese fishing boat

Battalion Landing Team: See BLT

Battalion Recon: USMC special warfare and reconnaissance unit under direct control of its division commander

Battery: An artillery unit consisting of three to six same-caliber cannons

Battery-Operated Grunt: Lingo for an Infantryman who carried the PRC-25 radio on his back

Battle Cross: See Field Cross

Battle Stations: Alarm for personnel to immediately get to a specific weapon or location; aka General Quarters

Battle: A violent, prolonged encounter between two large, organized armies, ships, or aircraft

Bayonet: A knife, blade, or spike fitted to a rifle muzzle and used to stab or slash an enemy

Bazooka: A 3.5-inch recoilless rocket launcher; used early in the war; replaced by the M72 LAAW; see also M1

BC: Medical notation meaning *Battle Casualty*

BCD: Bad Conduct Discharge; see addendums

BDA: See Bomb Damage Assessment

Be Advised: Initial statement that alerts listeners to imminent forthcoming important information

Bean Burner: Army cook

Beans & Balls: One of several C-Ration meals, officially *Beans w/Meat Balls in Tomato Sauce*

Beans & Dicks / Beans & Weenies: C-Ration meal, officially *Beans, w/Frankfurter Chunks in Tomato Sauce*

Beat Feet: Run away quickly

Beaucoup: French word for many or much or a lot or huge or very, etc.; aka Boo-Coo or Boo-Koo

Beehive Round: Rocket, shotgun shell, tank or artillery round, loaded with clusters of Flechettes

Belay: Navy term meaning fasten a rope to a cleat or other secure object

Belay: Navy term meaning stop or quit

Below: Any deck on a ship below the one you're on; under

Belt Suspenders: Supports a knapsack and the rifle belt with all its equipment

Benny Pack: See SP Pak

Berm: An artificial embankment of earth used as a defense against attack

Berth: A narrow bed fixed to the bulkhead of a ship

Berth: A ship's parking place at a pier or wharf

Betel Nut: Palm nut containing a mild narcotic; used to relieve tooth pain, but makes the mouth appear bloody

BFE: Bum Fuck Egypt; the place a Shitbird was threatened to be sent if he didn't square himself away

BGB: Big Gray Boat; vernacular for any ship that one can't identify

Bic: Vietnamese word meaning understand; *No bic* = not understood, *Khong biet* = Do not understand

Bid Whist: An African-American inspired card game that was a modified-rule, shortened version of bridge

Biere LaRue: The original Vietnamese "Tiger" beer, known for the tiger on its label; see also Tiger Piss

Big Boys, The: Lingo for artillery and tanks

Big PX: The United States; aka Land of the Big PX

Big Red One: Nickname for the 1st Infantry Division whose unit patch has on it a big red number 1

Big Red One: Satirical phrase attached to the division, *If you have to be one . . . be a Big Red One*

Bilge Rats: A derogatory Navy term for a ship's engine room sailors; aka Snipes

Billet: A job assignment or a residence

Bingo Fuel: Term indicating that an aircraft needs to return for fuel without delay, just before the PONR

Bipod: Two-legged, stanchion used to support rifles and guns

Bird Dog: Airborne Forward Air Controller; aka FAC

Bird Dog: O-1 Cessna *Bird Dog*; tandem seat, single-propeller, light, observation aircraft

Bird Shit: Derogatory term for paratroopers as they jump from an aircraft

Bird: Usually a helicopter, but can be any aircraft

Biscuit Bitches: Offensive, undeserved term referring to Donut Dollies

Bitchin: Very good or awesome

Bivouac: Temporary camp without tents or cover

BK: Casualty category term meaning *Below-the-Knee* traumatic amputation

Black Is Beautiful: A pride-fostering catchphrase referring to African-Americans

Black Ops: Black Operations; covert, top-secret activities

Black Syph: See Black Syphilis

Black Syphilis: Non-existent VD so contagious that if contracted, one would be listed as MIA and never go home

Blackbird: See SR-71

Bladder: Heavy, rubberized, collapsible container for petroleum or water; 500 to 50,000 gallons; aka Blivet

Blanket Party: Peer-driven corporal punishment for a person's infractions by flailing with soap bars in socks

Blast: Slang term meaning a great time, usually a party

Blinker Fluid: A fruitless-search prank on FNG's sent to find this non-existent liquid

Blitzed: Drunk on alcohol

Blivet: See Bladder

Blood Pack: Blood-volume expander intravenous field medical kit containing serum albumin

Blood Trail: A track of blood left by a dragged away dead man or by a fleeing man who has been wounded

Blood: Slang for an African-American

Blooper: See M79

Blousing Bands: Elastic ties used to secure uniform trouser cuffs to boot tops; aka blousing garters

Blow Away: Lingo for kill or destroy; see also or Grease or Hose or Nail or Waste or Zap

Blow Your Mind: Slang phrase meaning a thing or act that is amazing or unbelievable

Blow: Slang term meaning to waste something of value, usually money

BLT: A sea-going, reinforced Marine infantry task force ready for amphibious or helicopter insertion into combat

BLU: Bomb, Live Unit; see BLU-82

BLU-82: A 15,000 lb. above-ground-detonated bomb used to clear jungle landing zones; aka Daisy Cutter

Blue Feature: Any water such as a canal, river, lake, ocean, etc. on a combat map

Blue Line: A stream, river, or canal on a combat map

Blue Side: Navy term for those personnel not assigned to the Marine Corp; see also Green Side

Blue Water Navy: Naval forces on the ocean as opposed to those on rivers (Brown Water Navy)

BMFIC: Big Mother Fucker In Charge; generally the highest ranking officer in a division

BNR: Personnel record casualty suffix (usually after KIA) meaning *Body Not Recovered*

Boarded Out: After medical/mental panel (board) review, a person is separated from the military as unfit

Boat People: Refugees fleeing by boat

Body Bag: Zippered, heavy-duty, plastic bag for dead combat casualties; aka Glad Bag

Body Count: Number of dead bodies totaled after combat

Body Snatch: A Recon or SF act of stealthily abducting an enemy soldier from his own backyard for POW purposes

Bofors: Swedish autocannon firm; aka L/70; see 40mm Bofors

BOHICA: Bend Over, Here It Comes Again; refrain when some exasperating action is about to take place

Bomb Damage Assessment: LRRP/Ranger/Recon or air-photo review of large weapon target destruction; aka BDA

Bomb Farm: See Ammo Dump

Boneyard: Hue massacre burial site where 2,800-6,000 civilian non-combatants were mercilessly executed by NVA

Booby Trap: Improvised explosive or non-explosive injuring device; an automatic ambush

Boo-Coo, Boo-Koo: Bastardized French word; see Beaucoup

Booking: Slang for moving very quickly

Boom-Boom Girl: Slang for prostitute

Boom-Boom House: Slang for whorehouse

Boom-Boom: Slang for sexual intercourse

Boomer: Lingo for an AF technician operating the mid-air refueling arm on a KC-135 Stratotanker; aka Young Tiger

Boondocks: Rugged, untamed remote countryside; jungle; aka Boonies or Bush or Indian Country/Territory

Boondoggle: A pointless military operation poorly conceived

Boonie Hat: Soft, full-brimmed, cotton-poplin or rip-stop fabric, military hat with crown vents and a chin strap

Boonie Rat: A U.S. Army infantryman

Boonies: Rugged, untamed remote countryside; jungle; aka Boondocks or Bush or Indian Country/Territory

Boot Camp: Recruit training; Basic

Boot Louie: Lingo for a 2nd Lieutenant

Boot: A Marine recruit in training

Boot: Recruit in training; sometimes a brand new soldier just out of training; a guy new to a unit

BOP: Base Of Preference; Air Force Form 392 allowing 8 choices to request one's next assignment; aka Dream Sheet

BOQ: Bachelor Officer's Quarters; where officers live

Boss: A good or awesome thing

Bought The Farm: Got killed; references insurance money for the family to pay off their farm mortgage

Bouncing Betty: Mine that springs up to groin height before detonating

Bow: Front of a boat or ship

Bow-Wow: A derogatory term for someone in the Army; see also Dogface and Doggie

Box: Code word for an M-113 APC

Brain Housing Group: Drill Instructor term for a recruit's skull or head

Brass: Slang word meaning officers

Bravo Zulu: See BZ

Bread: Cash, money

Break Brush: To walk through the boonies (not on roads or trails)

Break Out: To take from its container or storage area

Break Squelch: To momentarily press a radio transmit button and not speak; generally a silent signal

Break Starch: The wearing of a freshly cleaned and starched work uniform

Breech Key: A fruitless-search prank on FNG's sent to find this non-existent artillery unlocking item

Brig Chaser: Navy term for a prisoner escort

Brig Rat: Navy term for an inmate or criminal detainee

Brig: Navy term for a jail; confinement area

Brigade: See addendums

Bring Smoke: To deliver overwhelming fire upon enemy positions

Bringing Pee: Descriptive term for the visual arc of thousands of rounds of tracers firing from a Spooky gunship

Bro: A non-derogatory term for a male African-American, a Brother

Broken Arrow: Radio code word to directly bombard a base or unit that is currently being overrun by the enemy

Bronco: A North American Rockwell OV-10; tandem-seat, twin-turboprop, 281mph, light attack/observation aircraft

Brother: A non-derogatory term for a Black male; aka Bro

Brothers-In-Arms: Term for the bond shared by all combat veterans

Brown Bagger: A married military man who usually returns home at night

Brown Bar: A second lieutenant/Navy ensign; the single gold bar indicating rank; see also Butter Bar

Brown Water Navy: Naval forces on rivers as opposed to those on the ocean (Blue Water Navy); see MRF

Browning Automatic Rifle: Used by ARVN and some Special Forces early in the war; see BAR

Bru: One of 54 tribes of Montagnard (mountain people) found in Vietnam

BS: Bullshit; a lie, something untrue

Buddy Plan: Recruit incentive wherein if you enlist with a friend, you are sure to go through Boot Camp together

BUF: Big Ugly Fucker; slang word for a Boeing B-52 Stratofortress strategic bomber

BUFF: Big Ugly Fat Fucker; slang word for a Boeing B-52 Stratofortress strategic bomber

Bug Juice: Insect repellent

Bug or Bugged: The act of annoying someone or of being annoyed

Bug Out: Leave quickly

Bulkhead: Navy lingo for wall

Bummed Out: State of depression; sad, dismayed

Bummer: A thing, act, or occurrence that depresses, disappoints, or frustrates; bad luck

Bunk: A bed or berth or rack

Bunker: A mostly underground, reinforced, fighting post or protective refuge from artillery, rockets, and mortars

Burning Shitters: The burning of sewage in 55-gal half drums taken from heads/latrines; see also Honey Dippers

Bush: Hostile territory outside the relative safety of a base or rear area; see also Boondocks or Boonies

Bust Caps: Fire a handheld weapon

Busted: To be found out or caught doing something one shouldn't, or to be arrested

Butt Can: Ashtray

Butt: A cigarette

Butter Bar: A second lieutenant/Navy ensign; a single gold bar indicating rank; see also Brown Bar

Butterfly: A guy who flits from female to female

Buttons: See Gravel Mines

BVR: Flight acronym for *Beyond Visual Range*

BX: Base Exchange; a store on an Army or Air Force installation; see also PX

By The Numbers: A series of steps in sequence; divided into small individual instructions

BZ: Well done; a pat on the back; aka Bravo Zulu

C

C & C: Command and Control; usually the helicopter used to manage and direct a battle by a commander

C. Turner Joy, USS: Destroyer DD-951 allegedly attacked by NVN; see Tonkin Gulf Incident and Maddox, USS

C.O. or CO: Commanding Officer

C.P. or CP: Command Post

C's: See C-Rations

C-123: Fairchild *Provider* twin-propeller, cargo/transport aircraft; aka "Dumbo"

C-130: Lockheed *Hercules* four-engine, military transport/cargo aircraft; aka Herky

C-135: A Boeing C-135 *Stratotanker*; a four turbojet, inflight refueling aircraft; aka Gas Station In The Sky

C-141: Lockheed *Starlifter*; strategic airlifter, four turbo-fanjet, long-range, transport/cargo aircraft

C-4: A soft, malleable, plastic chemical explosive; a small piece could be set afire to quickly heat C-Rats

C-47: Douglas "Skytrain" twin-engine, military transport/cargo aircraft; aka Gooney Bird

C-5A: Lockheed *Galaxy*; four turbofan, 532mph, 281,000lb cargo + 80 PAX, high-wing, heavy lift aircraft; aka FRED

C-7: Often misidentified pre-war version of the Caribou; actually the DHC-4 (not the C-7) was in Vietnam

CA: Combat Assault

Cache: A hidden stockpile of weapons, supplies, or food

Cadence: Rhythm or tempo in marching, often a sing-song chant

CAL or cal: Caliber

Call-Sign: Unique code word or words used to denote a person, unit, aircraft, or ship

Cam On: English phonic for the Vietnamese *Thank You;* (Cảm ơn – pronounced Kahm uhn)

Cammies: Camouflaged uniform

Can You Dig It?: Slang phrase for *Do you understand?* or *Do you like it?*

Candy Ass: A Non-Hacker or Ten-Percenter

Canister: Thin-skinned, short-range, disintegrating artillery round releasing large amounts of shrapnel without HE

Canned Rations: See C-Rations

Cannon Cockers: Affectionate term for artillerymen

Canteen: A portable water container (usually quart-size) made of aluminum or plastic or stainless steel

Canteen: An on-base convenience store/café

CAP: Combined Action Platoon; Marine unit living within a Vietnamese village to train and improve local life

Captain's Mast: NJP for minor offenses with no criminal proceeding required; see addendums

CAR: Combat Action Ribbon; Navy/Marine Corps version of the Army's Combat Infantryman Badge

CAR-15: XM177 Colt *Commando* select-fire, short-barrel, collapsible-stock, large-suppressor, variant of the M-16

Carbine: A shortened version of a standard rifle firing the same ammunition and usually used by SF troops

Care Package: A box of comfort items sent to a military man by family, friends, the Red Cross, or the USO

Caribou: See DHC-4

Carpet Bombing: When B-52's drop massive numbers of unguided bombs to destroy everything in a target area

Carry On: Resume the activity you were doing prior to being interrupted

Cartridge: Pre-assembled ammunition usually for a handgun, shotgun, or rifle

CAS: Casualty

CAS: Close Air Support; precision air strikes against an enemy very close to friendly units

Casual Company: A unit where Marines await individual reassignment

Casualty: A military person who is either wounded or dies in a war zone or is MIA or a POW

Cat: Slang for a cool dude, a hip male

Catch Ya On The Flipside: Slang phrase for *See you on my way back*

Cav or CAV: Short for *Cavalry* (armored or airmobile)

Cavalry: Originally soldiers highly mobile on horseback; in Vietnam soldiers were highly mobile in helicopters

CAVU: Aircraft code meaning *Ceiling And Visibility Unlimited*

Cayuse: See OH-6A

CB's: See SeaBees

CBU: Cluster Bomb Unit; a single bomblet; see Cluster Bomb

CBW: Chemical and Biological Warfare

CC: Command Chronology; documented report of significant events, sometimes minute-by-minute battle narrative

CC: Corrective Custody; aka Brig or Stockade or Correctional Custody or jail

CC: See Correctional Custody

CCB: Command & Control Boat; a river-borne radio relay

CCC: Command and Control, Central; Director of MACV-SOG operations central SVN. See also CCN and CCS

CCN: Command and Control, North; Director of MACV-SOG operations northern SVN. See also CCC and CCS

CCS: Command and Control, South; Director of MACV-SOG operations southern SVN. See also CCC and CCN

CE: Helicopter Crew Chief (Engineer)

Cease Fire: An official agreement between belligerent parties to stop fighting for a determined period of time

Cease Fire: An order to stop shooting immediately; see also Check Fire

Central Highlands: A 20,000 sq. mi. plateau area in central SVN of mostly hills, with mountains in the west

Cessna: See Bird Dog and O-1

CEV: Combat Engineer Vehicle; see M728

CG: Coast Guard

CG: Commanding General

CH: Combat Helicopter

CH-21: Piasecki *Shawnee*; single-radial engine, tandem rotor, 98mph, 265mi range, helicopter; aka Flying Banana

CH-3: Sikorsky *Jolly Green Giant,* heavy lift, single-rotor, rescue & recovery helicopter (USAF)

CH-34: Sikorsky *Seahorse* cargo/troop-transport, piston-engine, single-rotor helicopter (USMC & USN)

CH-37: Sikorsky *Mojave* heavy lift, cargo/troop-transport, twin piston-engine, single-rotor helicopter (USA & USMC)

CH-46: Boeing *Sea Knight* medium cargo/troop-transport, twin jet-engine, tandem-rotor helicopter (USMC & USN)

CH-47: Boeing *Chinook* heavy cargo/troop-transport, twin jet-engine, tandem-rotor helicopter (USA)

CH-53: Sikorsky *Sea Stallion* heavy cargo/troop-transport, twin jet-engine, single-rotor helicopter (USMC & USN)

CH-54: Sikorsky *Tarhe* heavy cargo, twin jet-engine, single-rotor helicopter; aka Sky Crane (USA)

Chaff: The radar-confusing and masking of bombers by ejecting strips of metal foil

Chain Of Command: See addendums

CHAOS: Sardonic acronym for *Commander Has Arrived On Scene*

Chaplain: A religious clergyman assigned to a branch of military service; see also Padre and Sky Pilot

Charge: Size or amount of explosive

Charlie: Slang for a decent enemy fighter; from military phonic "VC" Victor Charlie; see also Chuck and Mr. Charles

Charm School: Initial Army orientation/training for FNG's; see also Cherry School

Cheap Charlie: Vietnamese use of bastardized English for someone who spends little or no money

Check Fire: Radio code for artillery to stop firing immediately

Check It Out: Slang phrase for *Look at this*

Check Your Six: Look behind you; watch your back; sometimes used as a threat

Cherry School: Initial Army orientation/training for FNG's; see also Charm School

Cherry: Deriding term for a military person arriving In-Country for the first time; see also Fresh Meat and FNG

Cherry: Slang term for pristine; a hymen; a virgin; a person with no experience

Chick: Slang for a female

Chicken Plate: Protective chest armor for aircrew personnel

CHICOM or Chi-Com: Chinese Communist; usually a hand grenade, but can be anything made in Red China

Chief: Affectionate nickname for any military man of Native-American descent

Chiêu Hồi: Vietnamese for *Open Arms*; amnesty program for the enemy to defect to SVN; a VC or NVA defector

Chill: To calm down; relax

Chinook: See CH-47 helicopter; aka Shithook

Chit: A receipt or authorization written on paper

Chop-Chop: Slang for *hurry up*; food; to eat

Chopper: A helicopter

Chow Down: Eat

Chow Hall: An enclosed dining facility; aka Mess Hall or Mess Deck

Chow: Food; aka Mess

Christmas Bombings: 18-29 DEC 1972 bombardment of Hanoi and Haiphong that drove NVN to sue for peace

Christmas Bombings: Term dubbed by media to sway public opinion despite a 36hr bombing pause at Christmas

Chuck: Slang for a poor enemy fighter; see also Charlie and Mr. Charles

Chuck: A usually derogatory racial reference used by Blacks to denote a White person

Chuckerwood: A derogatory racial reference used by Blacks to denote a White person; aka Peckerwood

Chunker: Philco-Ford M75 40mm, 225rds per min, belt-fed, automatic grenade launcher on Huey & Cobra gunships

Church Key: A bottle opener

CIA: Central Intelligence Agency; aka The Agency or The Company

CIB: See Combat Infantryman Badge

CIC: Combat Information Center; aka Communications and Information Center

CIC: Commander-In-Chief; President of the United States

CID: Criminal Investigation Division; investigates crime in the military

CIDG: Civilian Irregular Defense Group; locals or tribesmen trained and led by U.S. advisers; aka Sidgee

CINCPAC or CinCPac: Commander In Chief, Pacific

CINCPACFLT or CinCPacFlt: Commander In Chief, Pacific Fleet

Cinderella Liberty: Time off ending at midnight

Civil Action Program: Strategic plan in which U.S. military personnel improve security and livelihood of villagers

Civvies: Civilian clothing

Clacker: An M57 Claymore electrical firing device; also used for detonating other explosive devices

Classes Of Supply: See addendums

Claymore: M18A1 above-ground, 60° fan-shaped, directional, anti-personnel mine w/700 ⅛" steel balls, 55yd range

Clearance: Military and/or political permission to engage the enemy in a specific area

Click: A single notch of adjustment on the sights of rifle; not to be confused with Klick

Clip: Device that holds ammunition cartridges in line to facilitate easy loading into a magazine

Close Air Support: Precision air strikes against an enemy very close to friendly units; aka CAS

Cluster Bomb: Munition air-dropped or ground-launched that scatters smaller bomblets (CBU's) over a large area

Cluster Fuck: Profane expression for when everything goes wrong

CMC: Commandant of the Marine Corps

CMH: *Congressional Medal of Honor*; America's highest military medal for heroism; aka MOH

Cô Vàn: Vietnamese word for military or civilian advisor assigned to train and mentor a SVN unit or political entity

CO: Commanding Officer

CO: Conscientious Objector; see Selective Service System in addendums

Coasties: Non-derogatory nickname for members of the United States Coast Guard

Cobra: Bell Aerospace AH-1; tandem seat, single-turbine, helicopter gunship; aka Huey Cobra and Snake

COC: Combat Operations Center; liaison between infantry, artillery, air, and other assets in a battle situation

Cocky-Dow: Bastardized Vietnamese word of unknown origin meaning kill or dead

COD: Entry on medical record meaning *Cause Of Death*

Code Of Conduct: List of six rules for U.S. servicemen to follow if they are taken prisoner; see addendums

Coka: Bastardized English word used by Vietnamese for *Coca-Cola*; pronounced *Coke-Ah*

Cold Cock: To punch someone by surprise

Cold Gas; Hot Gas: The refueling of an aircraft with the engine and/or rotors not operating (cold) or operating (hot)

Collateral Damage: Unintentional killing of civilians or destruction of non-enemy property

Collective: The rotor blade pitch control on a helicopter that causes the bird to climb/rise

Colors: Flag or identifying ensign; see also To The Colors

Column: When individuals or elements are placed one behind the other; aka File

Combat Engineer Vehicle: See M728

Combat Engineer: Expert in construction/demolition in support of friendlies while impeding the enemy

Combat Infantryman Badge: Military decoration for a soldier in ground combat, while assigned to an infantry unit

Combat Rations: See C-Rations

Combat Search & Rescue: See CSAR

Combat Skyspot: See GDB

Combat Tracker Team: Army term denoting a five-man/one-dog search group

Comm or Commo: Communications

Command Voice: Appropriately loud, authoritative-carrying, confident tone in one's speech

Commando Vault: Program name for using BLU-82 15,000 lb. bombs to clear landing zones in the jungle

Commando: Lightly-armed, elite soldier, specializing in hit-and-run raids

Commando: See V-100

Communications and Information Center: See CIC

Communism: Theory favoring all property and means of production be collective with no social or economic classes

Company, The: See CIA

Composition-4: See C-4

Compound: A fort or otherwise built-up military facility

ComUSMACV: Commander, U.S. Military Assistance Command, Vietnam

Concertina: Coiled razor wire used as a perimeter obstacle to deter an attacking enemy or escaping prisoners

Concussion Grenade: MK3A2 hand-thrown explosive device used to stun and disorient the enemy

Conex Container: Large, cube-shaped, steel shipping box used for transport and storage of military supplies

Confirmed: Term for a verified killing of the enemy

Cong: Communist; see also VC

Contact: Armed attack against or from the enemy

Containment: The suppression and control of communist expansion and influence by political and military means

Contested Area: The locale for which a battle is being waged

CONUS: Continental United States

Convoy: A group of vehicles (or ships) proceeding together usually escorted by armed troops and/or helicopters

Cook-Off: When a round self-fires from an overheated weapon

Cool It: To calm down or to stop what one is doing

Cool: Slang word for the state of being *with it* or popular; groovy

Coolie Hat: Wide, conical straw hat worn by Vietnamese for protection from the sun and rain

Cop Out: Slang for avoiding doing what one should rightly do; a weak excuse

Cordon & Search: See County Fair/Country Fair

CORDS: Civil Operations and Rural Development Support; a pacification effort coordinator

Cork: Nickname for the drug taken by LRRP's to prevent defecation while on covert patrols

Corps Tactical Zones: See addendums

Corps, The: Short non-derogatory term for The United States Marine Corps

Corpsman Up!: A desperate call for a Corpsman to come immediately to a wounded Marine

Corpsman: Navy enlisted-grade medical man serving in a Marine unit; aka Doc

Correctional Custody: Air Force term for a jail; confinement area; aka CC

Corsair: See A-7

Cosmoline: Sticky, grease-like, protective coating for storing and shipping weapons

COSVN: Central Office of South Vietnam; the Communist HQ for military and political action in South Vietnam

Cot: Portable, narrow, folding, wood-framed, canvas bed

Counterculture: Essentially, all the hippies and those who reject the values and norms of current civilized society

Counterinsurgency: Civilian and military efforts taken to defeat guerilla warfare and activities

Country Store: Any Vietnamese retail shop

County Fair/Country Fair: An operation to surround and search a village; aka Cordon & Search

Court-Martial: See Uniform Code Of Military Justice in addendums

Cover Me: *Shoot at the enemy to suppress their fire as I move to another location*

Cover My Ass / Cover Your Ass: Taking precautions to avoid detection or blame; aka CYA

Cover The Six: Protect or watch the rear

Cover: Any hat of a U.S. Marine

Cover: Concealment

Cowboy: Vietnamese thug; rider on a motorbike who swipes valuables from hapless individuals

Coxswain: The non-ship pilot/commander of a boat and its crew

CP: Command Post; a platoon or higher element's headquarters in the field

CQ: Charge of Quarters; a two-man guard/monitor of in-and-out traffic of a barracks or HQ overnight

Cracker Box: A field ambulance; the ¾-ton M-43 Dodge or the 1¼-ton M725 Kaiser

Crash & Dash: Slang for the act of an aircraft landing and then immediately taking off again; see also Touch & Go

Crash: Slang term meaning to fall asleep, generally from exhaustion

C-Rations: Individual canned meals; wet rations as opposed to dry rations; aka Combat Rations or Field Rations

C-Rats: Combat Rations

Crew Chief: The crewmember who maintains and assures the flyability of an aircraft

Crib: A person's home or place they sleep

Crispy Critters: Callous and emotion-avoidance term for burned-to-death enemy soldiers

Crosscheck: The act of examining and verifying that each other's gear is Good-To-Go

Crotch, The: Non-derogatory term for the Marine Corps when used amongst Marines

Crud, The: Warm-climate rashes, fungi, infections, boils, or ulcers affecting the skin; aka Jungle Rot or Trench Foot

Cruise: Navy term for period/length of enlistment; a tour of duty

Crusader: See F-8

CS: Oxidizing agent that temporarily burns mucous membranes of eyes, nose, and throat; aka Tear Gas or Riot Gas

CSAR: Combat Search And Rescue

CTZ: Corps Tactical Zone; see addendums

Củ Chi: Series of notorious, 3-level, 200km, interlocking tunnels dug by VC in which to hide and to launch attacks

Cumshaw: Navy term for unofficial pilfering, bartering, or scrounging for needed unavailable items; aka Scrounge

Cut Me A Huss: Phrase meaning *Do me a favor* or *Go easy on me*; see also Huss

CV-2: See DHC-4

CY: Calendar Year

CYA: Cover Your Ass; see also Cover My Ass

Cyclic: The rotor blade plane control on a helicopter that causes the bird to go forward, backward, left, or right

Cyclo: Three-wheeled bicycle taxi with a passenger seat in the front

D

DA: Department of the Army

DAF: Department of the Air Force

DAI: Casualty code for *Death from Accident or Injury*, a non-battle death

Daisy Chain: String of explosive devices usually hung over a trail

Daisy Cutter: BLU-82 15,000lb, and variants, air-burst bomb dropped by a C-130, used to clear jungle for an LZ

Dai-Uy (Die-wee): Vietnamese term for captain or village chief; (Đại úy)

Danger Close: Code term used to warn that air/artillery strike impacts will be perilously adjacent to your location

Dap: Custom by Black troops of shaking, tapping, and slapping, etc. of hands as a greeting or display of camaraderie

Dapsone: A once-a-week, anti-malarial/anti-leprosy, prophylactic tablet; aka Horse Pill

DARGAS: *Does Anyone Really Give A Shit?*

DD: Dishonorable Discharge; see addendums

DD: Navy *Destroyer*

DD-214: The release from active duty DOD form; aka Out Sheet

D-Day: The day that an operation is to be launched

DDG: Destroyer with guided missiles

Dead Marine Zone: Demilitarized Zone; sardonic term for the "neutral" area separating NVN from SVN; aka DMZ

Dear John: A letter from a girlfriend or wife back home declaring she is leaving you for someone else

Deck: Navy term for the floor

Decked Out: Slang for being attired in one's finest clothes

Decked: Knocked out by a fist during a fight

Dee-Dee Mau or Di-Di Mau: Bastardized Vietnamese word meaning to *Go away quickly!* or *Get out of here now!*

Dee-Dee or Di-di: Bastardized Vietnamese word meaning to *Go away!* or *Leave!*

Deep Serious: In considerable trouble

Deep Shit: In considerable trouble

Deep Six: Throw away; get rid of

Defilade: Natural obstacle, such as the side of a hill away from enemy observation or fire

Delta Tango: Designated Targets; preselected artillery targets for later bombardment

Delta, The: Short for Mekong Delta; IV-Corps; the southernmost Tactical Zone of SVN; see addendums

Demilitarized Zone: DMZ; "neutral" area separating North from South Vietnam; see also Dead Marine Zone

DentCAP: Dental Civilian Action Program; U.S. dental personnel attended to local's tooth problems and hygiene

Department Of Defense: Oversees all national security and military agencies and their functions; aka DOD

Deployment: Movement or relocation of forces and equipment to a favorable position for a military engagement

DEROS: Date of Estimated (or Earliest or Eligible or Expected) Return from Overseas

Desertion: The act of abandoning a post or duty with no intent to return; see also AWOL and UA

Desoto: U.S. Navy early code name for patrols in the Tonkin Gulf

Det Cord: Detonating Cord; long, ropelike, instantaneous fuse used to trigger an explosive, can be used to fell trees

Det: Short for Detachment

Detachment: Small unit capable of self-sustained operations

Deuce Gear: Marine slang for DOD Requisition Form 782; field equipment, usually web gear

Deuce Point: The second man in a patrol; aka Slack Man

Deuce: Two or second

Deuce-And-A-Half: A 2½ ton truck

Devil Dog: Term WWI Belleau Wood German soldiers labeled Marines for their fighting tenacity; see Teufelhunde

DHC-3: De Havilland *Otter*; twin-engine, propellered, STOL, utility transport aircraft; aka U-1A

DHC-4: De Havilland *Caribou*; single-engine, propellered, STOL, utility transport aircraft; aka CV-2

DI: USMC Drill Instructor

DIA: Defense Intelligence Agency; the military's covert operations and spy arm

Dibs: Childish term used to claim a right to something, such as first pick of a donut or a specific seat in a vehicle

Diddy Bag: Drawstring canvas bag for small items

Diddy-Bop: Slang for a carefree, careless swagger, instead of the military way of walking

Di-Di Mau: See Dee-dee Mau

Di-Di: See Dee-dee

DIE: Draft-Induced Enlistee; derisory term for men who enlisted to avoid looming conscription

Dien Bien Phu: Disastrous French defeat in 1954; a reminder to avoid bad planning or underestimating the enemy

Dien Cai Dau: Vietnamese for crazy; see also Dinky Dow (Điên cái đầu – literally *Crazy in the head*)

Dig: Slang term asking someone to pay attention to something; to like or enjoy something

Digger: Australian infantry soldier; aka Aussie

DILLIGAF: *Do I Look Like I Give A Fuck?*

Dime Nickel: A 105mm howitzer

Dink: Derogatory term referring to the enemy and sometimes all Vietnamese; see also Gook, Gooner, Slant, Slope

Dinky Dow: Bastardized Vietnamese word meaning crazy; see Dien Cai Dau

Direct Air Support Center: Central location for the coordination of air strike requests

Dishonorable Discharge: See addendums

Disneyland East: Facetious term for The Pentagon

Disneyland Far East: Facetious term for the U.S. Military Assistance Command, Vietnam (MACV) headquarters

Dispensary: Army clinic; see also Sick Bay

Div: Division

Dixie Station: Code name for U.S. Navy operational area in the South China Sea; see also Yankee Station

DMZ: Demilitarized Zone; "neutral" area separating North from South Vietnam; see also Dead Marine Zone

DN: Department of the Navy

Do Me A Solid: Phrase meaning to ask for a favor

DOA: *Dead On Arrival*

Doc: Affectionate nickname for a Corpsman or medic

DOC: Casualty code for *Death from Other Causes* such as homicide, suicide, etc.

DOD: Casualty code for *Died Of Disease* such as illness or natural causes

DOD: See Department Of Defense

Dog Tags: Stainless steel ID plates the size/look of animal tags; worn under the laces of jungle boots In-Country

Dog-And-Pony-Show: A military presentation for visiting dignitaries, politicos, or celebrities

Dogface: A derogatory term for someone in the Army; see also Doggie and Bow-Wow

Doggie Chocolate: See Tropical Hershey Bar or Montagnard Bar

Doggie: A derogatory term for someone in the Army; see also Dogface and Bow-Wow

Domino Theory: Cold War belief if one country falls to communism, so will its neighbors in relentless succession

Don't Know Shit From Shinola: Being ignorant or uniformed; NOTE: *Shinola* is a former shoe polish brand

Dong (Đồng): The basic Vietnamese monetary unit; see also Piaster

Don't Call Me Sir, I Work For A Living!: Usually declared by a non-officer who is mistakenly called *Sir*

Don't Flip Your Wig: Phrase meaning *don't get upset*

Don't Mean Nuthin': An emotion avoidance/coping term meaning something bad that happened isn't an issue

Don't Sweat It: Slang phrase meaning *don't let it get you down*

Donut Dollies: Morale-raising, young women, Red Cross volunteers who visited hospitals/bases and managed clubs

Door Gunner: Crewman who protects his side of a helicopter usually with an M-60 machinegun or a Ma Deuce

Dope: Intelligence information; see also Intel and Poop and Scoop and Skinny and Word

Dope: Marine term for the adjustments made to a weapon sight

Dope: Slang term for marijuana and other illicit drugs

Doper: Individual who uses marijuana and/or illicit drugs; see also Head

Double Time: At twice the pace in a formal march, similar to a trot; if not in a march, *hurry up*

Double-Digit Midget: A person with less than 100 days left in his tour of duty; see Short and Single-Digit Midget

Doubtfuls: Suspect indigenous persons not labelled Viet Cong or civil offenders

Dove: Slang for a person advocating placation and/or negotiation in a quest for peace; see also Hawk

DOW: *Died Of Wounds*; casualty code for when a wounded person dies after reaching a medical facility; see KIA

Downer: Slang for something depressing

Draft Dodger: A person who actively evaded military conscription by lying, faking, or moving to Canada

Draft, The: Slang for the actions of the Selective Service System; see addendums

Draftee: A non-voluntary person who is conscripted into military service; see Selective Service System

Drag Squad: Small unit that guards the rear behind the main advance element; aka Drag, The

Drag, The: See Drag Squad

Drag: Slang for something frustrating or disappointing or depressing

Dragon Wagon: M25 Tank Transporter

Dream Sheet: Slang for AF Form 392; see BOP

Dress Blues: Formal uniform worn by Marines

Dress Whites: Formal, lightweight uniform worn by Navy personnel

Drill Instructor: A Marine Corps boot camp NCO or SNCO trainer of recruits

Drill Sergeant: An Army basic training instructor-of-recruits, usually a Sergeant or Staff Sergeant

D-Ring: Handy D-shaped, snap clip used for attaching equipment to web gear and for repelling

Drip: A pus-like secretion from the penis indicating a sure sign you contracted gonorrhea

Drive-On Rag / Towel: A cloth, usually terrycloth, worn around the neck to wipe off sweat and for other purposes

DRNAV: Dead Reckoning Navigation; the use of visible points on the ground to pilot an aircraft

Drop A Dime: To call someone on a pay phone; to betray a confidence, to inform on someone

Drop And Give Me . . . : Basic training Drill Instructor's often punishment of assigning a certain number of pushups

DRV: Democratic Republic of (North) Vietnam; communists

Dry Fire: Practice; the exercise of firing an ammo-less weapon

Dry Rations: See LRRP Rations

Dry Run: Practice

DSA: District Senior Advisor; head person in a MAT

Duck Hook: Code name for the mining of Haiphong and other harbors & ports in July 1970; see also Pocket Money

Dud: A munition that has failed to detonate upon impact

Dude: Non-derogatory slang term meaning a male person

Duffle/Duffle Bag: A large oblong, canvas bag used to store and carry all of a soldier's uniforms; see also Seabag

Dumb Bomb: An unguided explosive munition dropped by aircraft; aka Iron Bomb or Gravity Bomb

Dumbo: Nickname for a C-123 Provider

Dump Truck: Slang word for a Boeing B-52 Stratofortress strategic bomber

Dung Lai: Vietnamese word for stop or halt

Dust Off: *Duty Uniform Services To Others Friend & Foe*; medical evacuation by helicopter

Dust Off: Radio code request for a UH-1 medevac helicopter w/no heavy guns and all but one row of seats removed

Duster: M42; an M24 tank chassis with an open turret and two 40mm AA guns converted to automatic cannons

Duty Hut: USMC drill instructor's office just off the squad bay

E

E&E: Escape and Evade; a predetermined, or spur-of-the-moment, survival route to avoid capture or death

Eagle Flight: Vietnamese airborne troops dropped in by helicopter to support ground troops in contact

Eagle Pull: Code name for the U.S. evacuation from Phnom Penh, Cambodia in 1975

Eagle Shits, The: Pay day

Early-out: As troop levels were being downsized, it was a reduction in the length of time to serve in the military

Easter Offensive: Failed 1972 NVA conventional warfare invasion of SVN utilizing 25 divisions and 700 tanks

Eat The Apple, Fuck The Corps: Amusing mantra between Marines; beware anyone else who utters those words

ECM: Electronic Countermeasures; jamming, detection, radar-deception, etc.

Egg Beater: Slang term for a helicopter

Ego Trip: The state of blissfully believing you are the center of the universe and that Your Shit Don't Stink

EH: Electronics Helicopter

Eighth & Eye: Slang for the oldest Marine Corps Post (at 8th and I in Washington, DC); home of the Commandant

Eight-One Mike-Mike: 81 millimeter; U.S. medium mortar

Eighty-One: U.S. medium mortar (81mm)

Eighty-Two: Communist medium mortar (82mm)

El T or El Tee: Informal/affectionate term for a Lieutenant; see also LT

ELD: Explosive Loading Detachment; USCG unit that supervised the off-loading of all munitions for the Army

Electric Strawberry: Nickname for the 25th Infantry Division's *Tropic Lightning* shoulder patch

Element: An individual, squad, platoon, or other section that is part of a larger unit; see addendums

Elephant Grass: Tall (up to 10'), sharp-edged grass

Elephant Turd: Amusing lingo for Blivet or Bladder

Elephant: Radio code for the M48 Patton tank

Eleven-Bravo: 11-B; Army infantryman MOS

ELINT: Electronic Intelligence

EM: Enlisted Man

Emergency Resupply: Ammunition and water only

Ensign: Entry-level officer in the Navy and Coast Guard, equivalent to a 2nd Lieutenant O-1; see addendums

Entrenching Tool: Short, folding shovel; see also E-Tool

EOD: Explosive Ordnance Disposal

ET: Electronics Technician

ET: Established Termination (of service); aka ETS

ETA: Estimated Time of Arrival

ETD: Estimated Time of Departure

E-Tool: M1967 tri-folding, short shovel used for digging and filling sandbags; locks folded or 90 or 180 degrees

E-Tool: See Entrenching Tool

ETS: *End Time of Service* or *Estimated Termination of Service*; the day one is out of the Army

Evac: Short for Evacuation Hospital; short for Medevac

Evac'd: Short for Evacuated, generally medically by helicopter

Every Swinging Dick: All persons in a unit

Exfil: Exfiltration; exiting from an area of operation

Expectants: A medical category term meaning wounded who are anticipated to die

Explosive Loading Detachment: See ELD

Extend: To volunteer for longer, or extra, time In-Country, usually a six-month tour of duty

Exterminate With Extreme Prejudice: To assassinate or kill at any cost

Extraction: Withdrawal, usually by helicopter

Eye Fuck: Drill instructor accusatory term when a recruit is looking where he should not look

Eyes In The Sky: Aerial reconnaissance

F

F-100: North American *Super Sabre;* single-seat/turbojet, Mach 1.4 supersonic, jet fighter-bomber; see also Hun

F-105: Republic *Thunderchief;* single-seat, single-turbojet, Mach 2.08 supersonic, fighter-bomber; aka Thud

F-4: McDonnell-Douglas *Phantom*; tandem-seat, twin-turbojet, Mach 2.2 supersonic, interceptor-fighter-bomber

F-8: Vought *Crusader*; single-seat, single turbojet, Mach 1.86 supersonic, carrier-based, fighter

FAC: Forward Air (or Artillery) Controller; an airborne (or ground unit) strike coordinator; aka Bird Dog

FAG: Field Artillery Group

Fag: Slang for a cigarette; also short for Faggot

Faggot: Slang term used derogatorily for a homosexual person; aka Fruit

Fair Force: Derisive term for Air Force females

Fall In: Gather together in a Rank and File formation

Fall Out: Break the rank and file formation, and then disperse

Fantail: Open area at the stern of a ship; see also Afterdeck

Far Out: Slang for *Wow!* or amazing

Fart Sack: A mattress cover or flight suit

Fast Mover: Jet aircraft, usually a fighter bomber; see also Slow Mover

Fat Body Platoon: Grueling USMC boot camp special element to make recruits quickly lose weight and get in shape

Fat Farm: Army version of Fat Body Platoon

Fatigues: The standard work/combat olive drab uniform for the U.S. Army

Favorite Turd: Affectionate term for a best friend

FB: See Firebase

FDC: Fire Direction Center; controls artillery adjustment calculations/impacts and warns infantry and aircraft

FE: Flight Engineer; crewmember who monitors and operates an aircraft's complex systems

Feather: The setting of propeller pitch to neither pull nor push air

Feet Dry: Pilot radio code meaning his aircraft is now over land

Feet Wet: Pilot radio code meaning his aircraft is now over open water

FFAR: Folding-Fin Aerial Rocket fitted with various warheads; see also Mk 4 and Zuni

Field Cross: Makeshift memorial of inverted rifle with helmet atop, dog tags draped, boots in front; aka Battle Cross

Field Day: Term used to denote an overall cleanup of quarters, squad bays, or barracks

Field Jacket: Medium-weight work coat worn over utilities or fatigues

Field Marching Pack: Marching Pack plus the Short Blanket Roll

Field Of Fire: An area in which a gunner, or group of gunners, will cover with their weapons

Field Rations: See C-Rations

Field Scarf: A necktie

Field Strip: To disassemble or take apart, usually a weapon

Field Surgical Kit: A small suturing/surgical set carried by Corpsmen and medics for emergency field surgery

Field Transport Pack: Transport Pack plus the Long Blanket Roll

Field, The: Hostile territory outside the relative safety of a base or rear area

Fi-Fi: *Fuck It, Forget It*

Fifty: Browning M2 .50-caliber heavy machinegun; 550 rounds per minute; aka Ma Deuce

Fifty-One: A 12.7mm communist heavy machinegun; aka 51mm

Fighting Hole: A temporary protective hole dug by troops, usually as a night defensive position; aka Foxhole

FIGMO: Finally (or Fuckit), I Got My Orders; a feeling of blissful abandon when receiving orders out of Vietnam

File: See Column

Fini-Flight: A pilot's last mission

Fire Direction Center: Controls artillery adjustment calculations for correct impact of rounds; see also FDC

Fire Fan: A Field Of Fire for artillery or mortars

Fire Fly: One helicopter flies lights on and a second lights out and kills whatever shoots at the first helo; see Snoopy

Fire For Effect: Firing all artillery at same time at a common target continuously until ordered to check-fire

Fire In The Hole!: A shouted warning of an imminent detonation of explosives

Fire Mission: A radio request or assignment to fire artillery, mortars, or naval guns at an enemy target

Fire Support Base: Usually a temporary base for artillery/mortars to bombard the enemy in an operation; aka FSB

Fire Watch: Basic training overnight Sentry chore shared by recruits in 2-hour increments to get used to Guard Duty

Firebase: Artillery/mortar unit(s) usually on elevated terrain and secured by infantry; see also FB

Firecracker: M449 artillery round clustered with 60 bouncing golfball-size bomblets which detonate mid-air

Firefight: A violent exchange in which small-arms weaponry is fired between opposing elements; see also Skirmish

Firefly: A searchlight-equipped helicopter; aka Lightning Bug

First Aid: Emergency treatment of wounded or injured persons before medical personnel or a medevac arrives

First Light: The time at which the sun is 12-degrees below the horizon; morning twilight

First Shirt: Slang for *First Sergeant*; see also Top and addendums

Five O'clock Follies: Mocking term used by frustrated reporters for the daily press briefings by MACV in Saigon

Fix: A location

Fixed Wing: Non-helicopter aircraft

Flak (or Flack) Jacket: Ballistic nylon/fiberglass-paneled vest for protection from shrapnel and subsonic munitions

Flakey: Slang term for mental untidiness; eccentric; peculiar

Flame Drop: 55-gal drums of Foo-Gas (Flame Fougasse) helicopter-dropped and ignited mid-air over enemy bunkers

Flamethrower: Any variant of fire-propelling weapons; see M2-2 and M67 and M132 and PGM-5 and Zippo

Flaming Asshole: An assuredly ridiculous person

Flaming Asshole: The firing of a supersonic jet aircraft's afterburners

Flank: The extreme right or left side of a group of troops

Flank: The movement of troops to confront the side elements of the enemy making him divide his forces and focus

Flankers Out!: The order for specific assigned elements to move to, and protect, the sides of the advancing unit

Flare: Artificial illumination delivered to light up a battle area by a hand-fired device, mortar, artillery, or aircraft

FLC: Force Logistics Command; unit responsible for equipping Marines with arms, food, uniforms, and supplies

Flechette: Small, steel dart fired in a cluster from a shotgun, rocket, artillery or tank round; see also Beehive Round

Flick: Slang for a movie

Flight Engineer: Crewmember who monitors and operates an aircraft's complex systems; see also FE

Flight Line, 50 Feet of: A fruitless-search prank on FNG's sent to find this non-existent rope

Flight Line: Airfield area for parking and for hangers, where aircraft are maintained and serviced

Flip Out: Slang term meaning to lose control of one's emotions, usually in a ranting state

Float Phase: A sea Deployment for a Marine unit

Float: A Deployment or Tour of Duty on a ship

Flower Child: Slang for a person who wore floral-themed decorations to symbolize peace and love; see also Hippie

Flower Seeker: Slang for a man searching for a prostitute

Fly Right: To be straightforward and honest in all things

Flyboy: Mocking term for anyone in the Air Force, especially a pilot

Flying Banana: See CH-21

Flying Boxcar: See AC-119

Flying Ladder: See Simmons Rig and Jacob's Ladder

FMF: Fleet Marine Force

FMFPAC: Fleet Marine Force, Pacific

FNG: Fucking New Guy; a military person arriving In-Country for the first time; see also Cherry and Fresh Meat

FO: Forward Observer; infantry-attached artillery officer who calls in artillery, air, and/or Naval gunfire

FOD: Medical record code for *Finding Of Death*, after 1 year without a body a person may be deemed to be dead

FOD: Foreign Object Damage to an aircraft such as bird strikes, hail, stones, dirt, objects left on the runway, etc.

Foo-Gas: Flame Fougasse; usually in base-perimeter drums that eject burning napalm during an enemy attack

Force Logistics Command: See FLC

Force Recon: USMC special warfare and reconnaissance unit under direct control of the FMF commander

Foreign Object Damage: See FOD

Forward Air Controller: An airborne (or on-ground) air strike/artillery coordinator; see FAC

Forward Observer: Men trained in the art of calling in artillery and all other forms of heavy fire at enemy targets

Forward: Navy term meaning toward the front of the ship or boat

Fougasse: Ancient mortar-like, burning-explosive, weapon system; see Foo-gas

Four-Deuce: 4.2" (107mm) M30 rifled-bore, heavy mortar with a range of 4mi; aka

Fox Four: Radio code for a McDonnell-Douglas F-4 *Phantom*

Foxhole: See Fighting Hole

Foxy: Slang for sexy

FPO: Fleet Post Office; all mail sent to Navy/Marine personnel in Vietnam went via FPO, San Francisco; see also APO

Fracture Jaw: Top secret plan to move nuclear weapons into SVN for use against the North at a moment's notice

Frag: A short (fragmentary), informal combat order

Frag: A fragmentation grenade

Fragging: The act of murdering a military person by his fellow troops; usually with a fragmentation grenade

Frankenstein: A booby trap consisting of C-4 wrapped in a spool of barbed wire

Freak Out: Slang meaning to lose control over one's emotions; to have a panic attack; become wildly irrational

Freak: Generally a slang term meaning a Hippie

Freaky: Slang for an eerie event or thing, something quite inexplicable or bizarre

FRED: *Fucking Ridiculous Economic/Environmental Disaster*; C-5A aircraft so-named for its massive fuel appetite

Free Fire Zone: An area where permission to fire at a target is not required; aka Free Strike Zone

Free Love: Unrealistic notion wherein sex with anyone, any time, without obligation or emotional bond, is feasible

Free Strike Zone: See Free Fire Zone

Freedom Bird: Any aircraft that would fly a military person out of Vietnam, usually to the U.S.

Freedom Train: Code for retaliatory strikes below the 20th Parallel due to the NVA's 1972 Easter Offensive invasion

Freq or Freak: Radio frequency

Frequent Wind: Code name for the U.S. evacuation from Saigon in 1975

Fresh Meat: Slang for a military person arriving In-Country for the first time; see also Cherry and FNG

Friendlies: Non-enemy allied personnel

Friendly Fire: Bombardment/shooting mistakenly directed at U.S. or allied troops; see also Misadventure

Fruit Salad: The rows of colorful ribbons/badges worn on dress uniforms

Fruit: Derogatory term for homosexual; aka Faggot

FSA: Forward Support Activity; a logistical unit augmenting an operation

FSB: See Fire Support Base

FTA: Derogatory term meaning *Fuck The Army*, usually written as acronym graffiti

FTA: Fire The Artillery

FTA: Ironic term meaning *Fun, Travel, Adventure*

FUBAR: Derogatory slang meaning *Fucked Up Beyond All Recognition (Repair)*; an impossible thing

Fuck You Lizard: A Tokay Gecko known for its loud, nighttime mating call, *faaaaa-cue!*

Fucked Up: Badly wounded or drunk or stoned or something done that was foolish/stupid

Fuckin' A: Fucking Affirmative; strongly agreeing or strongly stating

Fugazi: Slang word meaning screwed up; completely out of whack

FULRO: *United Front for the Liberation of Oppressed Races*; Montagnard struggle for a separate tribal State in VN

Funky: Slang for something good or something bad, depending on context

Funny Money: See MPC

Funny Papers, The: Ironic term for colorful combat maps

Fuse / Fuze: An electrical or mechanical mechanism that detonates to trigger a much larger explosive

Fuze – Barometric: A fuse that detonates at a predetermined elevation above sea level

Fuze – Cord: A rope-like fuse that burns at a precise rate after ignition

Fuze – Impact: A fuse that detonates immediately or near-immediately after colliding with a target

Fuze – Timed: A fuse when triggered, detonates after a predetermined amount of time

Fuze – VT: Variable-Time; a proximity fuse that detonates at a predetermined distance from a target

FW: Casualty code for *Fragmentation Wound*

FX: Casualty code for *Fracture*

FY: Fiscal Year

G

Gaggle: A flying formation of helicopters

Galaxy: See C-5A and FRED

Galley: Navy term for kitchen

Gangway!: *Move aside! Get out of the way!*

Gangway: A narrow walkway, usually on a ship

Garand: M1; semi-automatic, World War II-vintage standard infantry rifle; saw limited use in the Vietnam War

Gas Mask: M17A1 Field Protective Gas Mask; respirator protecting against chemical and biological weapons

Gas Station In The Sky: A Boeing C-135 *Stratotanker*; inflight refueling aircraft

Gas: Any activity invoking joy; fun

GCA: Ground Control Approach; wherein ground-based radar in inclement weather "talks in" a pilot to a runway

GD: General Discharge; see addendums

GDB: Ground-Directed Bombing; ground unit guiding of B-52 target course correction/bomb release; see Skyspot

GDHC: General Discharge Under Honorable Circumstances; see addendums

Gear: Equipment

Gedunk / Geedunk: A store that sells snack foods; the snack food itself

General Discharge: See addendums

General Orders: Like the 10 Commandments, but there are 11 for the military; see addendums

General Quarters: Alarm for personnel to immediately get to a specific weapon or location; aka Battle Stations / GQ

Geneva Accords: Truce between French and Viet Minh ending the Indochina War and establishing the 17th Parallel

Get All Your Ducks In A Row: Slang phrase meaning to get organized

Get All Your Shit In One Dixie Cup: Slang phrase meaning to get organized and to get rid of all unnecessary items

Get Down!: Slang for *Start dancing!*

Get It On: Slang phrase meaning fight, or to have physical romance

Get It Together: Slang phrase for *Calm down* and *Straighten up your act*

Get Real: Slang phrase for *Stop being naïve*

Get Some!: Excited expletive in the heat of battle referring to the killing of the enemy

Get Your Shit Together: Slang phrase for *Pack your gear*; *Shape up mentally*; an order to do much better

GI Bill: Benefits package for honorably discharged veterans

GI Can: Garbage can

GI House: A place where garbage is kept

GI Joe: A non-derogatory nickname for a U.S. Army soldier; aka Joe

GI Party: Cleaning of the Barracks

GI: Government Issue

GI: Slang term meaning soldier

Gig Line: Imaginary straight line formed when the tie, shirt buttons, belt buckle, and zipper are in perfect alignment

Gimme Some Skin: Slang phrase meaning a salutatory request for a handshake or a Dap or a palm slide

Gizmo: Slang term for an indescribable part or gadget

Glad Bag: Irreverent, emotional-avoidance term for a Body Bag

GM: Gunner's Mate

Go To Hell Hat: Bush hat of Thai-based, Aussie Thud pilots, missions marked red for Hanoi/black elsewhere in NVN

God Botherers: Slang term for Chaplains

GOFO: *Grasp Of the Fucking Obvious*

Go-Go Boots: Knee-high, colorful, shiny boots usually worn with Mini-Skirts

Gold Star Mother / Father: A parent whose son or daughter has died in the military

Golden BB: The one improbable small bullet that shot down the helicopter or other aircraft

Gomer: An enemy soldier

Good-To-Go: The point at which a person, unit, or equipment is ready to be put into action

Gook: Derogatory term referring to the enemy and sometimes all Vietnamese; see also Dink, Gooner, Slant, Slope

Gooks In the Wire!: Verbal alarm warning that enemy soldiers are attempting to infiltrate the perimeter

Gooner: Derogatory term referring to the enemy and sometimes all Vietnamese; see also Dink, Gook, Slant, Slope

Gooney Bird: See C-47

Gourd: Slang term for your head; brain

GP: General Purpose

GQ: General Quarters; alarm ordering personnel immediately to a specific weapon or location; aka Battle Stations

GR: See Graves Registration

Grab-Ass: Slang for playful or frivolous activities

Grab Their Belts To Fight Them: Enemy tactic of closing-in to reduce the effect of massive U.S. supporting arms

Grass: Slang for marijuana; see also Dope or Weed

Gravel Mines: Small, air-dropped, antipersonnel mines released in clusters of 1,400-7,000 at a time; aka Buttons

Graves Registration: Unit for identification, remains repatriation preparation, and personal effects custody of KIA's

Gravity Bomb: An unguided, explosive munition dropped by aircraft; aka Dumb Bomb or Iron Bomb

Grease Gun: M3A1 Submachine Gun; 6" barreled, .45cal, 400 round-per-minute, close-quarters, automatic rifle

Grease: Lingo for kill or destroy; see also Blow Away or Hose or Nail or Waste or Zap

Greaser: Derogatory term for a person of Latin-American or Mediterranean descent

Greasers: A subculture of greasy-haired toughs who wore leather jackets, cuffed jeans, and ankle-high boots

Great Mistakes: Derogatory nickname for the *Great Lakes Naval Training Center*

Green Beanies: Disparaging term for the U.S. Army's Green Beret soldiers

Green Beret: SF soldier cross-trained in combat arms, counterinsurgency, language, airborne, LRRP, covert actions

Green Eye: See Starlight Scope

Green Machine: Affectionate term for The United States Marine Corps

Green Side: Navy term for USN personnel (usually Corpsmen) assigned to the Marine Corps; see also Blue Side

Green Tracer: A visible color streak left momentarily in the air to track the path of a bullet (communist); see Tracer

Green Weenie: Derisive term for the United States Army

Greens: Army Class-A winter uniform

Grenade: A short-range, hand- or rifle- or launcher-projected, chemical-releasing smoke or explosive weapon

Grenadier: The man whose primary weapon is the M79 grenade launcher

Grid Coordinates: Set of letters & numbers matching those on a map for a precise location

Grid Square: The 1,000 meter x 1,000 meter quadrangle represented by horizontal and perpendicular map lines

Grid Squares, A Box of: A fruitless-search prank on FNG's to find this non-existent item

Grinder: See Parade Ground

Groovin': Slang for enjoying something

Groovy: Slang term for something pleasing or wonderful

Ground Fire: Non-artillery shooting targeted at aircraft

Ground Pounder: The basic Infantryman; a Grunt

Group Grope: Sardonic term for a substantial formation of helicopters

Grunt: A slang term for Marine infantryman, usually E-1 to E-5; see also Ground Pounder and Infantryman

GRUNT: Sarcastic phrase meaning *General Replacement Usually Not Trained*

GS: Gunship

GSW: A casualty category referring to *Gun Shot Wound*

GSW-T&T / GSW-TTH: A casualty code referring to *Gun Shot Wound, Thru & Thru*

G-T Line: Gun-Target Line; the straight path covered by a fired round

Guard Channel: Radio emergency frequency that is heard by all aircraft no matter which channel they have on

Guard Duty: Assignment of protecting or watching over personnel, buildings, equipment, or enemy approaches

Guerrilla Warfare: Stealthy, unconventional fighting techniques employed generally by indigenous forces

Guerrilla: Organized resistance soldier using stealthy, unconventional fighting techniques

Guidon: Small flag or pennant displaying a unit's designation

Gulf of Tonkin Resolution: 1964 law giving the U.S. President power to wage full war in SEA with no war declaration

Gun Truck: A cargo vehicle with added armor and machineguns that provides convoy security

Gun: A long-barreled artillery cannon, either self-propelled or towed or mounted on a ship

Gung Ho: Bastardized Chinese slogan for *Working Together*; enthusiastic; motivated; committed

Gunny: Affectionate term for a Marine Corps *Gunnery Sergeant*; see addendums

Guns: Affectionate term for a Marine Corps *Gunnery Sergeant*; see addendums

Gunship: A heavily-weaponized attack helicopter or other aircraft that can deliver substantial firepower on a target

GVN: Government of (South) Vietnam

Gyrene: Slang term meaning Marine

H

H & I: Harassment and Interdiction; the random firing of artillery into enemy territory

H&HC: Headquarters and Headquarters Company; a separate non-line element at battalion or higher level

H&MS: Headquarters & Maintenance Squadron; administration and maintenance unit of a Marine Aircraft Group

H're: One of 54 tribes of Montagnard (mountain people) found in Vietnam

Hắc Báo: Elite *Black Cat* ARVN unit specializing in extractions of downed aircraft and crews in Laos

HAC: Helicopter Aircraft Commander

Hải Vãn Pass: Treacherous Highway One gap over the Truong Son mountain range just north of Da Nang

Halazone: A chlorine-based tablet used to disinfect one quart of drinking water

Half-Track: M3 multi-use, 10 ton, light-armored, 45mph, 200mi range, front-wheeled/rear-tracked vehicle

HALO: High Altitude, Low Opening; parachuting of men or the dropping of supplies from above SAM or AAA range

Ham & Mothers / Ham & Motherfuckers: A C-Ration meal of gelatinized, minced ham and lima beans

HAM: Hairy-Assed Marine; derogatory term for a male Marine; see also BAM

Hamlet: A small village

Hammer & Anvil: The driving out of an enemy unit in the direction of a waiting annihilation force

Hand Crank: Tool used to elevate and traverse an artillery tube when hydraulics fail

Hang A Light: Radio call for flares to be fired overhead

Hang Loose: Standby; relax

Hang Up: Slang for a focused emotional problem or phobia

Hanoi Hannah: A well-known, enemy female radio broadcaster who played popular tunes and spread propaganda

Hanoi Hilton: Ironic nickname for the notorious Hoa Lo Prison where American POW's were confined and tortured

Hanoi Jane: Scornful nickname for actress Jane Fonda for going to NVN to provide aid and comfort to the enemy

Hanoi: Capital city of NVN; Hà Nội

Happy Ending: The sex act when the massage is over wherein the male ejaculates; see Steam & Cream

Harassment & Interdiction: See H&I

Harbor Site: SF/SEAL/Recon night hunker-down area, with heavy-vegetated concealment and easily defended

Hard Charger: Lingo for a motivated military person

Hardhat: A slang term meaning construction worker

Hassle: Lingo for *to nag* or *to annoy someone*; a chore that is a nuisance

Hatch: Navy term for a door or doorway

Hatchet Force: U.S./CIDG combined recon team assigned to reinforce an actively-compromised A-Team

Haversack: Lightweight marching pack with its own suspender straps

Hawk, The: Slang term for cold, windy weather

Hawk: A U.S. SAM; see LAAM

Hawk: Slang for a person advocating vigorous action, or war, instead of appeasement; see also Dove

Hawser: A thick, heavy cable or line used to moor or tow a ship

HCM: Ho Chi Minh, the leader of NVN

HD: Honorable Discharge; see addendums

HE: High Explosive

Head Call: Navy lingo for using the bathroom or toilet

Head Man: Person in charge of a local area, usually a village chief

Head Motherfucker In Charge: The highest-level commander in a unit or government entity or village; aka HMFIC

Head Shed: Slang for Headquarters

Head: Individual who uses marijuana and/or illicit drugs; see also Doper and Pot Head

Head: Navy lingo for bathroom or toilet; see also Latrine

Headquarters: The place or offices from which management, control, and administration of a command takes place

Hearts & Minds: Pacification program to win support of the Vietnamese by political, economic, and social means

Heat Tab: Small, rectangular, flammable Trioxane tablet used for heating C-rations

HEAT: High Explosive Anti-Tank; a fin-stabilized armor piercing missile

Heavy: Slang term meaning intellectually deep; emotional weight; sad feeling

Hedge Row: Dense line of vegetation that grows at the edges of fields, paddies, or village homes; aka Tree Line

Helgoland: German hospital ship based in the Han River, Da Nang, that treated Vietnamese

Helmet Liner: Six-point suspension system w/adjustable bands and chin strap attached to the inside of a Steel Pot

Helmet: Protective armored head covering; aka Steel Pot

Hercules: See C-130

Herky: Nickname for a C-130 Lockheed *Hercules*

Hero: Derogatory reference for a person who is taking stupid chances just to get a medal; aka John Wayne

Hershey Squirts: Slang term for diarrhea

HES: Hamlet Evaluation System; analysis of required monthly computerized lists of questions, sent in by DSA's

HH-43F: See Pedro

HH-53: See Super Jolly Green Giant

HHC: See H&HC

H-Hour: The time that an operation is to be launched on D-Day

High & Tight: A shaved-side, very short on top Marine haircut

High Angle Fire: The trajectory of mortar or artillery where the projectile arcs higher than its distance downrange

High Explosive: Instantaneous-reacting chemical that turns to gas so abruptly as to cause a violent/shattering effect

Highway One: The major north-south route from Hanoi through Saigon and into the Mekong Delta

Hill Fights: Specifically refers to the April-May 1967 DMZ battles for Hills 861, 861Alpha, 881South, and 881North

Hill: Any rising elevation above sea level (hills and mountains) marked in meters on a map

Hip: Slang term for aware and knowledgeable of something

Hippie: Slang for a person typically with long hair and flowery outfits, often promoting drug use and peace and love

Hit The Deck: *Dive for cover*; also *get up and go to work*

Hit The Head: Lingo for go to the bathroom

HM: Hospitalman; MOS of a Corpsman

HMFIC: See Head Motherfucker In Charge

HMH: Helicopter, Marine Heavy; a CH-37 or CH-53

HML: Helicopter, Marine Light; UH-1E

HMM: Helicopter, Marine Medium; CH-34 or CH-46

Hmong: One of 54 tribes of Montagnard (mountain people) found in Vietnam

Ho Chi Minh Sandals: Footwear utilizing old tires cut into soles and held on feet with straps cut from inner tubes

Ho Chi Minh Trail: The enemy supply route of jungle roads, trails, and paths through Laos and Cambodia into SVN

Ho Chi Minh: The leader of North Vietnam; aka HCM and Uncle Ho

Hog: M60; 7.62mm, belt fed, air-cooled, 600 rounds per minute, light machinegun; aka Pig or M2 or Ma Deuce

Hồi Chánh: An enemy soldier who defected to SVN under the Chieu Hoi amnesty program

Hollywood Marine: A teasing reference for a Marine who endured San Diego boot camp as opposed to Parris Island

Honcho: Bastardized Japanese word (Hanchō) meaning Head Man; the boss; the person in charge

Honey Dippers: Individuals on a Honey Patrol detail who stir the burning shit in Honey Pots until it's all gone

Honey Patrol: The act of gathering latrine excrement barrels (honey pots) to a central area for burning

Honey Pot: Half of a 55-gallon drum placed under a latrine or head to catch and hold human excrement

Honey Wagon: A low vehicle or trailer used to gather Honey Pots

Honky: African-American contemptuous term for a White person

Honorable Discharge: See addendums

Hooch: See Hootch

Hook: Shortened slang of Shithook for a CH-47 Chinook helicopter

Hook: The communications radio or its handset; aka Horn

Hootch Girl: A vetted Vietnamese woman hired to do menial jobs such as cleaning SEAhuts and washing clothes

Hootch: Shelter or residence, usually a SEAhut; a simple, rural Vietnamese home

Horn: The communications radio or its handset; aka Hook

Horse Pill: See Dapsone

Hose Down: To apply massive automatic weapons fire on an enemy target, usually from an aircraft

Hose: Lingo for kill or destroy; see also Blow Away or Grease or Nail or Waste or Zap

Hot Gas; Cold Gas: The refueling of an aircraft with the engine and/or rotors operating (hot) or not operating (cold)

Hot Hoist: The helicopter extraction of a person using a cable and winch while under fire by the enemy

Hot LZ: A landing zone currently under fire by the enemy

Hot Toc: Vietnamese word for a barbershop; hớt tóc

Hot: A place currently under fire, or an area of likely ambush

Hotel Charlie: Phonic alphabet phrase for *Hot Chow*

House Mouse: Usually a diminutive recruit that acts as a Drill Instructor's aide for menial tasks

Housewife: Lingo for a sewing kit

Howard Johnsons: Sardonic term for any Vietnamese food pushcart street vendor

Howitzer: A cannon, either self-propelled or towed, that is capable of high or low angle firing

HQ: See Headquarters

Huế: Imperial capital city of Vietnam from 1802 to 1945; site of the 1968 Tet Offensive 26-day urban warfare battle

Huey Cobra: An AH-1 Cobra helicopter gunship; aka Cobra or Snake

Huey Slick: Bell UH-1 *Iroquois* series lightly-armed, utility helicopter used for medevac and to carry troops

Huey: Affectionate nickname for the Bell UH-1 *Iroquois* series of helicopters, and the iconic image of Vietnam

Hump: To walk on patrol; to be burdened with carrying something

Hun: Nickname for the F-100 fighter-bomber; "Hun" refers to hundred

Hurry Up And Wait: Military habit to get somewhere hastily, or prepare quickly, and then dawdle about

HUS-1: Sikorsky *Seahorse*, the forerunner to the CH-34, asked by name when Marines were in trouble; see HUSS

Huskie: See HH-43F

Huss: Originated from HUS-1 meaning *a favor* or *a break* or *help me out*; see also Cut Me A Huss

I

I Heard That: Slang phrase used to acknowledge a statement and be in agreement

I Shackle: Radio prefix words enunciated prior to sending encrypted grid coordinates

I Shit You Not: Slang phrase meaning a person is telling the truth or is very serious

I&I: Intoxication & Intercourse; derogatory term mocking the term R&R (Rest & Recreation)

Ia Drang Valley Battle: First major U.S. Army ground combat operation in SVN (14-18 November 1965)

IA: Immediate Action is necessary

IBGB: Itty Bitty Gook Boat; derogatory term for a very small Vietnamese watercraft; see also LBGB and LFGB

I-Corps: Northernmost of the four military regions in South Vietnam; see addendums

Id10t Form: A fruitless-search prank on FNG's sent to find this non-existent document

Idiot Stick: A rifle

IFR: Instrument Flight Rules; SOP for flying blind w/o reference to the horizon or ground; aka Popeye; see also VFR

IG: Inspector General

II-Corps: Next lower of the four military regions in South Vietnam below I-Corps; see addendums

III MAF: Third Marine Amphibious Force; consisting of the HQ Command of all U.S. Marines units in SVN

III-Corps: Next lower of the four military regions in South Vietnam below II-Corps; see addendums

IKE: Nickname for Dwight D. Eisenhower, 34th President of the United States

Illum or Illumination: A flare to light up a battle area delivered by a hand-fired device, mortar, artillery or aircraft

Image Intensification Device: See Starlite Scope

Immersion Foot: Caused by prolonged wetness resulting in open ulcers, fungal infection, etc.; see also Paddy Foot

Improvise, Adapt, and Overcome: Marine mantra for this routinely under-funded service to do more with less

In The Field: Any area outside of a camp where one can expect to encounter the enemy

In The Rear With The Beer / Gear: Infantryman's antipathy towards the REMF and the Pogue

Inactive Status: Period of time after uniformed service awaiting final discharge from the armed forces

Incident: Reportable occurrence or any enemy contact

Incoming: Verbal warning that enemy artillery, mortar, or rocket fire is, or is about to, start impacting your position

In-Country R&R Centers: China Beach for I-Corps; Nha Trang for II-Corps; and Vung Tau for III- and IV-Corps

In-Country R&R: 3-day break from war for grunts deemed deserving, not assured – based on merit; see also R&R

In-Country: In Vietnam

Incursion: A sudden and brief cross-border raid

Indian Country: See Indian Territory

Indian Territory: Countryside where the enemy has inhabited and is operating; aka Boondocks/Boonies/Bush

Indians: The enemy

Indig / Indigenous: Indigenous people; native to an area

Infantryman: A land-combat individual who fights on foot rather than from a vehicle, boat, or aircraft; aka Grunt

Ink Stick: Lingo for a writing pen

Insert: Insertion; deliver, usually Recon Marines, Green Beret's, SEAL's, or LRRP's to an AO point of entrance

Insertion: Delivery of Recon Marines, Green Berets, SEAL's, or LRRP's, to an AO point of entrance

Intel: Intelligence information; see also Dope and Poop and Scoop and Skinny and Word

Intelligence: Information deemed valuable of a military or political nature to further the goals of a mission or policy

Interdiction: The act of stopping or disrupting the movements and/or supplies of a looming enemy force

Interval: A prescribed distance between each person, especially in a column of infantrymen

Intruder: See A-6

IP: Instructor Pilot

IPW: Interrogation, Prisoners of War; a joint U.S./SVN program for the questioning of POW's; see also ITT

IR-20 Rice: First strain of high-yield *Miracle Rice* introduced to Vietnam in the 1970's

IR-22 Rice: Improved strain of high-yield *Miracle Rice* introduced to Vietnam in the 1970's

IR-5 Rice: First strain of high-yield *Miracle Rice* introduced to Vietnam in the 1960's

IR-8 Rice: Improved strain of high-yield *Miracle Rice* introduced to Vietnam in the 1960's

Irish Pennant: Lingo for a thread dangling from a uniform indicative of an unkempt appearance

Iron Bomb: An unguided explosive munition dropped by aircraft; aka Dumb Bomb or Gravity Bomb

Iron Hand: Mission code name to suppress NVN's SAM missiles and anti-aircraft artillery

Iron Triangle: Hotly-contested 120sqmi area between the Saigon and Tinh Rivers; included the Cu Chi tunnel system

Iroquois: See UH-1

Irregulars: An armed, non-standard military member who utilizes unconventional tactics

IRRI: International Rice Research Institute; developers of Miracle Rice

ITO: Instrument Take Off; using dash gauges/ground control rather than visual cues to maintain a runway heading

ITR: Infantry Training Regiment; the advanced combat training immediately after Marine Corps boot camp

ITT: Interrogation-Translator Team; see also IPW

IV-Corps: Southernmost of the four military regions in South Vietnam; see addendums

J

Jacob's Ladder: See Simmon's Rig and *Genesis 28: 10-13*

JAG: Judge Advocate General's Corps; handles all aspects of military justice and law

Jar Of Bulkhead Remover: A fruitless-search prank on FNG's sent to find this non-existent item

Jarai: One of 54 tribes of Montagnard (mountain people) found in Vietnam

Jarhead: Slang term for a U.S. Marine

Jazzed: Slang term for excited

JCS: Joint Chiefs of Staff; team of the top leader of each branch of the armed services, located at the Pentagon

Jeep: Any variant of a ¼-ton, 4WD, off-road capable, light utility vehicle

JEEP: Satirical term for *Just Enlisted, Expecting Promotion*

Jerry Can: A 5-gallon, pressed steel container used primarily for gasoline

Jesus Nut: Large nut that holds the main rotor to the helicopter transmission driveshaft

Jet Jockey: Mildly-affectionate term for a jet aircraft pilot; aka Zoomie

JFK: John Fitzgerald Kennedy; 35th President of the United States

Jink: Flight maneuver consisting of violent and unpredictable direction changes to evade enemy gunners

Jody: The invented name for a guy who is supposedly dating your wife/girlfriend while you are training or oversees

Joe Shit the Ragman: A sloppy, repugnant, unenviable individual; boyfriend of Rosie Rottencrotch or her sister Suzy

Joe: A non-derogatory nickname for a U.S. Army soldier; lingo for GI Joe

Joe: Coffee

John Wayne Can Opener: See P-38

John Wayne It: The act of pushing through pain, fear, and exhaustion like that Hollywood hero would

John Wayne Move: An action of a person that places himself in grave danger either heroically or foolishly

John Wayne Peanut Butter: C-Ration rock-hard peanut butter in a can

John Wayne: Mocking term for a person, or his actions, who exposes himself unnecessarily to danger

Jolly Green Giant: HH-3 Sikorsky heavy, long-range, armed, CSAR helicopter; not a Super Jolly Green Giant

JP-4: Jet Propellant; aviation fuel

JROTC: Jr. Reserve Officer Training Corps; prep program for high school students to become officers; see also ROTC

Juicer: A person known to be a heavy drinker of alcohol beverages

Jungle Boots: Light, nylon-canvas-leather, combat footwear with steel, sole inserts and ventilation/drain eyelets

Jungle Bunny: Derogatory term used by Whites toward Blacks

Jungle Fatigues: Army tropical camouflage combat uniform

Jungle Penetrator: Teardrop-shaped device with fold-down seats used to extract personnel from the jungle

Jungle Rot: Warm-climate rashes, fungi, infections, boils, or ulcers affecting the skin; aka The Crud or Trench Foot

Jungle Utilities: Marine Corps tropical camouflage combat uniform

Junk On The Bunk: All military clothing precisely laid out on the Rack for inspection; aka Things On The Springs

Junkie: Slang term for a person addicted to heroin

K

K: A kilometer; aka Klick

K9: Refers to a military dog and its handler

Kak Wheel: A KAL-55B coordinate-encryption disk tool carried by an RTO

KAL-55B: A Kak Wheel

KBA: Casualty code meaning *Killed By Artillery*

K-Bar / KA-BAR: U.S. Marine Corps Mark 2, *Knife, Fighting Utility*; 12" (7" blade) edged combat weapon

KBV: Casualty code meaning *Killed By Aircraft*

KC-135: See Stratotanker

Keystone: President Nixon's program to incrementally phase out U.S. troops and turn over the war to SVN

KHA: Killed, Hostile Action; casualty classification usually meaning the death of a non-combatant

Khaki: Tan, sandy color, used in some light, tropical or summer uniforms

Khmer Rouge: Cambodian communist forces

KIA: Killed In Action; casualty code when a person dies outright or dies before reaching a medical facility; see DOW

KIA-BNR: Casualty code meaning *Killed In Action, Body Not Recovered*

Kibosh: Slang word meaning to stop what is being done or to stop what is being planned

Kick Out: Term referring to the act of throwing out supplies at low altitude from a helicopter unable to land

Kill (or Killing) Zone: Ambush vicinity where it is expected that all enemy soldiers within will die or be wounded

Kill Ratio: Any statistical comparison gauging the total of something as it relates to the goal of dead enemy bodies

Kill Zone: Maximum effective radius of an explosive munition that will kill at least 95% of what is caught in that area

Killer Junior: Defense tactic using 105 or 155 artillery HE rounds set to airburst at 30 ft over attacking enemy troops

Killer Senior: Defense tactic using 8" howitzer HE rounds set to airburst at 30 ft over attacking enemy troops

Killer: Slang term meaning an amazing or truly impressive act; something extraordinary

Kiowa: See OH-58

Kit Carson Scouts: Enemy defectors who serve as scouts, interpreters, and intelligence agents for SVN and the U.S.

Kitchen Patrol: Work in a mess hall; aka KP

Kiwis: Lingo for New Zealanders

Klick: Vernacular for kilometer; aka K; not to be confused with Click

Knapsack: Transport and field pack attached to Belt Suspenders

Knock It Off: Slang for stop whatever it is you are doing

Knot: Nautical speed; 1 knot per hour = 1.151mph

Knuckle Dragger: Derogatory term for an aircraft mechanic

KP: Work in a mess hall; aka Kitchen Patrol

L

L/70: See 40mm Bofors

Laager: An Army term meaning to *Circle the Wagons*; a defensive circle of weaponry, usually at night

LAAM: Light Antiaircraft Missile unit; *Hawk* mobile missile system firing the MIM-23 SAM

LAAW: M72 Light Anti-Armor Weapon; single-round, 66mm, rocket launcher, disposable; see also LAW

Ladder / Ladderway / Ladderwell: Navy lingo for stairs

Lai Dai: Bastardized phonic of Vietnamese for *come here,* lại đây

Laid Back: Slang for someone or something that is relaxed and easygoing

Lam Son 719: Ill-fated 1971 ARVN incursion into Laos, Lam Sơn

Land Line: Telephone communications

Land Of The Big PX: The United States

Landing Craft: Any shallow-draft boat designed specifically to transfer troops and equipment from sea to shore

Landing Mat: PSP-type 22" x 12' runway-forming material without holes which helps to eliminate helicopter FOD

Landing Platform, Dock: See LPD

Landing Platform, Helicopter: See LPH

Landing Ship, Dock: See LSD

Landing Ship, Tank: See LST

Landing Zone: Any place a helicopter can land; aka LZ

Lanyard Grease: A fruitless-search prank on FNG's sent to find non-existent lanyard (cord) grease (lubrication)

Later: Slang for goodbye

Latrine: A bathroom; toilet; see also Head

LAW: M72 Light Anti-Tank Weapon; single-round, 66mm, rocket launcher, disposable; see also LAAW

Lay Chilly: Slang for to be motionless; *do not move*

Lay It On Me: Slang phrase for *speak your mind*; *tell me what's going on*; *give it to me*

Lazy Dog: An air-launched cluster munition approximately .50cal units, finned, kinetic energy projectiles

LBE: Load Bearing Equipment; web gear

LBGB: Little Bitty Gook Boat; derogatory term for a very small Vietnamese watercraft; see also IBGB and LFGB

LBJ: Long Binh Jail; a notorious U.S. Army stockade in Long Binh, SVN

LBJ: Lyndon Baines Johnson, 36th President of the United States

LCM-6: See Mike Boat

Lead Sled: See F-105

Lead: The first aircraft in a flight of two or more

Leaf Killers: Affectionate nickname for Dumbo pilots who spray defoliants

Lean, Mean Fighting Machines: Label for scrawny Marines exiting the bush due to a meager box-a-day C-Rat diet

Leatherneck Square: A quadrangle formed by Con Thien, Gio Linh, Dong Ha, and Cam Lo in northern I-Corps

Leatherneck: Nickname for a U.S. Marine; refers to the leather neck collar worn when sword fighting on ships

Leave: Authorized absence away from the military, 30 days per year (accrued at 2½ days per month)

Left-Handed Monkey Wrench: A fruitless-search prank on FNG's sent to find this non-existent tool

Leg: Army term meaning non-airborne infantryman; see also Straight Leg

LFGB: Little Fucking Gook Boat; derogatory term for a very small Vietnamese watercraft; see also IBGB and LBGB

LGB: Laser-Guided Bomb

LIB: Light Infantry Brigade

Liberty: A no more than 96-hour authorized absence, not chargeable to yearly leave

Lieutenant Without A Map: Mockingly, the second most dangerous thing in the world

Lieutenant With A Map: Mockingly, the most dangerous thing in the world

Lifer: A derogatory nickname for people who make the military a career

LIFER: Derisive term meaning *Lazy Inefficient Fucker Evading Reality* or *Lazy Inefficient Fucker Expecting Retirement*

Light `Em If You Got `Em: Take a cigarette break

Light `Em Up: To fire a weapon, or weapons, on the enemy

Lightning Bug: A searchlight-equipped helicopter; aka Firefly

Lima Charlie: See Loud & Clear

Lima Delta: Line of Departure

Lima Papa: Listening Post; aka LP

Lima-Lima: Aircraft radio phrase meaning *Low Level*

Lima-Lima: Land Line (telephone)

Limited Political War: Conflict to attain certain partisan goals without committing all existing resources and assets

Line Of Departure: The step-off point designated to begin a coordinated military operation

Linebacker I: Code name for Apr-Oct 1972 bombing campaign of NVN in response to their Easter Offensive invasion

Linebacker II: Code name for the renewed December 1972 bombing of NVN; dubbed *Christmas Bombings* by media

Lip Flappin: Slang for spreading gossip; talking just to talk

Liquid Sunshine: Ironic term for rain

Listen Up: Quiet down and pay attention to what I have to say

Listening Post: See LP

Lit Up: Past tense of Light `Em Up

Litter: Stretcher on which to carry the wounded or sick; see also Poncho Litter

Little People: Radio code for any Vietnamese, but usually ARVN's

LLDB: Luc Luong Dac Biet; SVN's Special Forces, patterned after the U.S. Green Berets, Lực Lượng Đặc Biệt

LMG: Light Machinegun

Loach: Nickname for a Hughes OH-6A *Cayuse* Light Observation Helicopter; aka LOH

Load Toad: A derogatory term for a weapons loader; aka Weapons Weenie

Loaded: Drunk or otherwise intoxicated

LOC: Line Of Communication

Local: A Vietnamese civilian

Lock & Load: To set the safety and chamber a round in a rifle or automatic pistol

Log Bird: A helicopter carrying supplies

Log: Logistics

LOH: Light Observation Helicopter; see Loach

Lollygagging / Lollygagger: Being lazy, goofing off, not pitching in, doing something not useful

Long Binh: Snide remark referring to a prostitute; also the name of a town and a U.S. base northeast of Saigon

Long Blanket Roll: Everything in the Short Blanket Roll plus another blanket

Long Green Line: The column of Marines who came before you since November 10, 1775

Long Range Reconnaissance Patrol: See LRRP

Long Rats / Long Rations: See LRRP Rations

Long Time / Short Time: Time length for which a prostitute is hired and figures heavily in payment negotiation

LORAN: Long-range Radio Navigation; course-plotting system using two or more fixed, pulse-type radio stations

Loud & Clear: Response to a radioed question about how the operator can be heard, if well

Love You Long Time / Love You Short Time: Prostitute's offer which figures heavily in payment negotiation

LP: Listening Post; nighttime hole or hide for 2 or 3 troops outside their unit's perimeter to act as early warning

LPD: Landing Platform, Dock; a ship supporting a small number of helicopters, amphibious craft, and combat troops

LPH: Landing Platform, Helicopter; a small carrier converted (in the 1960's) to support a helicopter squadron at sea

LRRP Rations: Lightweight, freeze-dried food for Long Range Reconnaissance Patrols; aka Long Rats or LRRP Rats

LRRP Rats: Lingo for lightweight, freeze-dried food for Long Range Reconnaissance Patrols

LRRP: Long Range Reconnaissance Patrol; nickname for an individual member of a LRRP; see also Lurp

LSA: Logistics Support Activity (Area): depot/yard/warehouse complex/hub where pre-distribution supplies are kept

LSA: Lubricant, Small Arms

LSD: Landing Ship, Dock; an amphibious craft used in ship-to-shore assault operations and as a transport

LSD: Lysergic Acid Diethylamide; an illicit and powerful hallucinogenic drug; aka Acid

LSMR: Landing Ship, Medium, Rocket; 207' ship with equivalent firepower of 5 destroyers; aka Rocket Rainmaker

LST: Landing Ship, Tank; an amphibious craft that can deliver tanks, vehicles, cargo, and troops onto shore

LT: Informal, affectionate term for a Lieutenant; see also El T and the addendums

Luke The Gook: A sardonic term for an enemy soldier; Luke for short

Lurp: Lingo for an individual member of a LRRP team; dehydrated food packs eaten by LRRPs

LVT: Landing Vehicle, Tracked; versatile amphibious craft with 5 variants In-Country; see also AmTrac

LZ Watcher: Enemy sentinel who reports to his superiors any activity in his assigned LZ to observe

LZ: Landing Zone

M

M / m: Meter, as a measurement; one meter = 3.281 feet

M: Morphine

M1 and M2: Carbine variants, World War II-vintage Garand rifle

M1: A 3.5" recoilless rocket launcher; aka bazooka; used early in the war; replaced by the M72 LAAW

M1: See Helmet

M1: Semi-automatic, World War II-vintage standard infantry rifle; Garand

M101: Rock Island Arsenal, towed, 2.5 ton, aluminum-carriaged, 105mm Howitzer; 7.0 mi range

M102: Rock Island Arsenal, towed, 1.5 ton, aluminum-carriaged, 105mm Howitzer 9.4 mi range

M106: FMC Mortar Carrier; 12.9 ton, M-113 APC chassis, tracked vehicle, with a traverse-mount 4.2" M30 mortar

M107: FMC 175mm, 28 ton, 50mph, self-propelled artillery gun, with a firing range of 25 miles

M108: Cadillac 105mm, 21 ton, 35mph, self-propelled howitzer, with a firing range of 9 miles

M109: 155mm, 27.5 ton, 35mph, self-propelled howitzer, with a firing range of 19 miles

M110: GMC 8-inch (203mm), 18.3 ton, 30mph, self-propelled howitzer, with a firing range of 14 miles

M113: Food Machinery Co. Armored Personnel Carrier (APC); 13.6 ton, 42mph, tracked, aluminum combat vehicle

M114: Cadillac 7.5 ton, 36mph, low-silhouette, tracked, aluminum, armored Command and Reconnaissance vehicle

M115: 8-inch (203mm), 16 ton, towed, heavy gun, with a firing range of 20 miles

M117: Un-guided, air-dropped, 750lb bomb

M125: FMC Mortar Carrier; 12.9 ton, M-113 APC chassis, tracked vehicle, with a traverse-mount 81mm M29 mortar

M125: Mack heavy, 10-ton, diesel-powered, 42-mph, cargo truck

M132: FMC 12 ton, 43mph, tracked, aluminum, flamethrower vehicle, with a firing range of 186 yards; see Zippo

M134: See Minigun

M14: See Toe Popper

M14: Springfield .30cal., 7.62mm, standard U.S. infantry semi-automatic battle rifle which replaced the M1 Garand

M151: See MUTT

M16: Colt Manufacturing .223cal, standard U.S. infantry automatic assault rifle which replaced the M14

M17A1: See Gas Mask

M18: A throwable signaling/screening, smoke grenade with red, orange, yellow, green, blue, violet or black smoke

M18A1: See Claymore

M1911A1: Colt .45 automatic, 7-round magazine, service pistol

M1921: See Thompson or Tommy Gun

M1941: See Marine Corps Pack

M1956: Canvas, load-bearing equipment such as pistol belt, pack, first aid case, canteen cover, etc.; aka Web Gear

M1967: See E-Tool

M1A1: See Bangalore

M2: A 60mm smoothbore, light infantry mortar with a range of 1.1 miles

M2: Browning .50-caliber heavy machinegun; 550 rounds per minute; aka Ma Deuce

M203: A Colt Manufacturing 40mm single-shot, grenade launcher fitted under the barrel of an M16 rifle; see SPIW

M2-2: Portable, 70lb, back-carried, 1 pressure tank + 1 napalm tank; five 2-second bursts, 50yd flamethrower gun

M24: Cadillac 75mm, 20 ton, 35mph, self-propelled, light reconnaissance tank

M25: Tank Transporter; aka Dragon Wagon

M-26: A U.S. 16oz, lemon-shaped, serrated-coil fragmentation, hand grenade

M274: See MULE

M29: An 81mm high-angle, infantry heavy mortar

M2A1-7: A portable, backpack flamethrower with an effective range of 44 yards

M-3: Medical Aid Bag, Small; tactical, casualty-size medical kit used by Corpsmen/medics; see also Unit One / M-5

M3: See Half-Track

M30: See Four-Deuce

M34: White Phosphorus, 27oz, hand grenade

M3A1: See Grease Gun

M40: Watervliet Arsenal 106mm, breach-loaded, single-shot, recoilless rifle

M42: See Duster

M422: See Mighty Mite

M43: See Cracker Box

M449: See Firecracker

M48: See Patton Main Battle Tank

M49A1: See Trip Flare

M5: Bayonet for an M1 Garrand

M-5: Medical Aid Bag, Large; tactical, platoon-size, medical kit used by Corpsmen/medics; see also Unit One / M-3

M50A1: See Ontos

M548: FMC 13 ton, 38mph, tracked APC chassis, cargo (usually ammo) vehicle

M55: See Quad-50

M551: See Sheridan

M56: Cadillac 90mm *Scorpion,* 7 ton, 28mph, tracked, self-propelled, unarmored, anti-tank gun

M57: See Clacker

M578: FMC 27 ton, 37mph, boom-craned, tracked, light recovery vehicle used for light, armored vehicle salvage

M6: Bayonet for an M14

M60: 7.62x51mm crew-served, belt-fed, 600 rounds per minute, air-cooled, light machinegun; see also Hog or Pig

M60: Chrysler 105mm, 47 ton, 30mph, "Patton-design," main battle tank

M60: See AVLB

M61: See Vulcan

M67: Detroit Arsenal flamethrower M48 Patton tank conversion, with a firing range of 300 yards; see also Zippo

M67: See Baseball Grenade

M7: Bayonet for an M16

M706: See V-100

M72: See LAAW and LAW

M725: See Cracker Box

M728: Chrysler 50 ton, 30mph, Combat Engineer Vehicle, tracked, self-propelled, mine-clearing, breaching, tank

M75: See Chunker

M79: Springfield 40mm, single-shot, break-action/breach-loaded, 400m, grenade launcher; aka Thumper or Blooper

M82: Un-guided, air-dropped, 500lb bomb

M88: A 51 ton, 30mph, boom-craned, tracked, heavy recovery vehicle used for tank/heavy equipment salvage

Ma Deuce: Browning M2 .50-caliber heavy machinegun; 550 rounds per minute; aka 50 or The 50

MA: Mechanical Ambush; a boobytrap

MAAG: Military Assistance Advisory Group; conventional warfare training mentors and military aid logistics

MAB: Marine Amphibious Brigade; see also MEB

MABS: Marine Airbase Squadron, the housekeeping unit of a MAG

Mac The FAC: Affectionate nickname for a Forward Air Controller

MAC: Military Airlift Command; long-range, strategic airlift of troops and cargo from the U.S. to the Pacific

Mach: The speed of sound (+/- 767mph); the measurement of velocity of a supersonic aircraft

MAC-SOG: Military Assistance Command, Studies and Observation Group; covert counterinsurgency warfare unit

MACV: Military Assistance Command, Vietnam; U.S. HQ responsible for the overall conduct of the war

MACV-CORDS: Civil Operations and Rural Development; U.S. and SVN pacification plan to gain support of locals

Mad Minute: A frustration-release act of firing weapons on full-automatic for one minute; aka Mike-Mike

Maddox, USS: Destroyer DD-731 attacked with torpedoes and machineguns by NVN; see Tonkin Gulf Incident

MAF: Marine Amphibious Force; see III MAF

MAG: Marine Air Group

Mag: Short for ammo Magazine

Magazine: An ammunition and explosives storage site, usually a super-strong, attack-resistant, half-buried structure

Magazine: Spring-loaded canister that holds and feeds ammunition into a weapon

Maggie's Drawers: A red flag or disk waved over the target at a shooting range to indicate a miss; a missed shot

Maggot: Drill Instructor's demeaning term for a boot camp trainee

MAGTF: Marine Air-Ground Task Force; the basic structure for all mission-specific, combined-arms missions

Mail Call: Much looked-forward-to part of the day (or week) when incoming mail is distributed

Main Force: Organized, standardized military units with identical uniforms and equipment

Make A Hole!: *Get out of the way!*

Make Love Not War: Idealist anti-conflict slogan

Make Love And War: Marine Corps catchphrase for doing both

Make Your Buddy Smile: Tighten up the queue

Malaria: Tropical, mosquito-borne disease, causing fever, exhaustion, vomiting, and headache, deadly if untreated

Malayan Gate: Tripwire-actuated booby trap made of a taught, spiked bamboo shaft; aka Bamboo Whip

MAM: Military-Age Male; initial classification for a detained suspect who is not under arms

Mamasan: Bastardized Japanese word meaning honorable mother; an older Vietnamese woman

Man, The: Slang term for a person of authority or a government entity

Marching Pack: Haversack, belt suspenders, and rifle belt combination; see Field Marching Pack

MARDIV: Marine Division; the 1st MARDIV and 3rd MARDIV served In-Country

Marine Corps Pack: Combination of the Haversack, Knapsack, and Belt Suspenders

Marine: A member of the United States Marine Corps; never referred to as Soldier

Mark Time: To march in place without covering any ground; to bide one's time

Mark-15 Retarding Device: Bomb pop-out fins that slow its descent to allow the plane time to escape the explosion

Marker Round: Single round revealing its impact with smoke or fire to adjust for barrage; aka Spotter/Target Round

Market Time: The U.S. and SVN Navy's campaign to stop the flow of enemy troops and materials by sea into SVN

MARS: Military Affiliate Radio Station; civilian ham radio operator program to relay radio to telephone calls home

Marston Mat: See PSP

Marvin The ARVN: A sardonic term for a member of the South Vietnamese Army

Mas-Cal: Classification term for *Mass Casualties*

MASH: Mobile Army Surgical Hospital; two of these units served in Vietnam

Mast: Non-Judicial Punishment for minor offenses with no criminal proceeding required; aka Office Hours

MAT: Mobile Advisory Team; small group of military mentors who lived with, and trained, ARVN at their outposts

MATS: Maybe Again, Tomorrow, Sometime; sarcastic nickname for Military Air Transport Service

MATS: Military Air Transport Service; since 1948 moving troops and supplies; superseded by MAC in 1966

Matty Mattel: Contemptuous term for the hated early M16 variant which often failed, refers to plastic stock/grips

MAW: Marine Air Wing; only the 1st MAW served In-Country

MAWS: Missile Approach Warning System; alert to an aircrew that a SAM has been fired at them; see SAM Song

Max Ord: See Maximum Ordinate

Maximum Consternation: Gen. Westmoreland quote meaning the creation of pandemonium

Maximum Ordinate: Highest trajectory point in flight of an artillery, mortar, or naval gun projectile; aka Max Ord

MBT: Main Battle Tank

MCAS: Marine Corps Air Station

McGuire Rig: SF 120' rope-hung extraction device with a painful Swiss Seat, later rigged with a swing-like seat

McNamara Line: Planned high-tech and choke-point barrier system from Tonkin Gulf to Laos just below the DMZ

MCRD: Marine Corps Recruit Depot; there are two, Parris Island, SC and San Diego, CA

Meal, Combat, Individual: Official term for a C-Ration repast

Meat Factory: A mordant, emotion-avoidance term for hospital

Meat Grinder, The: See Quad-50

Meat Wagon: A mordant, emotion-avoidance term for any medical emergency vehicle

MEB: Marine Expeditionary Brigade; a 6,000-man force-in-readiness consisting of 3 BLT's and a HQ group

Mech: Short for Mechanized infantry

MEDCAP: Medical Civil Action Program; village visits by U.S. medical personnel to treat disease and ease suffering

Medevac: Medical evacuation by helicopter; see also Dust Off

Medic: Enlisted-grade medical person

Medical Aid Bag, Large: See M-5

Medical Aid Bag, Small: See M-3

Mekong Delta: Tactical Zone IV; alluvial plain formed by divergent branches of the Mekong River; see also The Delta

Mellow: Slang term meaning without harshness; easygoing, relaxed, placid

Menu: Air strikes in Cambodia code names of Breakfast, Lunch, Dinner, Snack, and Dessert; see Operation Menu

Mermite Can: Portable, aluminum, insulated, hot or cold, food container

Mess Deck: A Mess Hall on a ship

Mess Hall: An enclosed dining facility; aka Chow Hall or Mess Deck

Mess: Food; aka Chow

Messmen: Helpers in a Mess Hall or Mess Deck

Met Message: Meteorological station-prepared weather report

MFIC: Motherfucker In Charge

MFW: Wound code meaning Multiple Fragmentation Wounds

MG: Machinegun

MIA: Missing In Action

Midnight Requisition: Theft; steal something

Mid-Rations: Food served for on-duty personnel between midnight and 06:00; aka Mid-Rats

Mid-Rats: See Mid-Rations

MIG: Mikoyan-Gurevich (MiG) Soviet fighter-bomber, jet aircraft; subsonic MiG-17; supersonic MiG 19 & 21

MIGCAP: MIG Combat Air Patrol; an aircraft mission specifically to hunt down and kill MIGs; see MIG

Mighty Mite: AMC ¼-ton, 1,800lb, 62mph, 4WD, Marine Corps tactical truck (jeep); aka M422; not Mity Mite

Mike Boat: LCM-6 Landing Craft, Mechanized; shallow-draft troop/supply boat

Mike Force: CIDG quick-reaction team

Mike: Minute in military phonics

Mike: See Mad Minute

Mike-Mike: Millimeter in military phonics

Miliaria: Small, pimplelike, stinging bumps that erupt when sweat cannot escape the skin; aka Prickly Heat

Military Air Transport Service: Multi-service mover of men and supplies until replaced with MAC in early 1966

Military Airlift Command: See MAC

Military Payment Certificate: See MPC

Military Phonic Alphabet: See addendums

Military Region 1: I-Corps; see addendums

Military Region 2: II-Corps; see addendums

Military Region 3: III-Corps; see addendums

Military Region 4: IV-Corps; see addendums

Military Training Instructor: Air Force Staff NCO's who train recruits; aka TI

Military-Industrial Complex: Notion that ties between the military and civilian defense industry is what causes war

Milk Run: An easy vehicle or aircraft mission where no hostilities or problems are expected

Million-Dollar Wound: A wound serious enough to be sent home, but not permanently crippling

MIM-23: Hawk medium-range SAM; solid-fuel rocket engine, 1,290lb, 16ft-8in x 14.5dia, Mach 2.4, 65K ceiling

Mind Game: The attempt, using flawed but seemingly rational arguments, to control the way another thinks or acts

Mine Magnet: Universal nickname for an AmTrac, APC, tank, or most any vulnerable armored vehicle

Mini Frag: Netherlands-built V40, 4.8oz, 1.6-inch round, Special Forces mini-fragmentation hand grenade

Minigun: General Electric M134, 30cal, 7.62mm, rotating, six-barreled gun, firing rate of 6,000 rounds per minute

Mini-Skirt: An unusually short and revealing skirt; a juxtaposition vying women's liberation with that of a sex object

Miracle Rice: High-yield rice plant introduced to Vietnam; see IR-5 or IR-8 or IR-20 or IR-22 and IRRI

Misadventure: Category of casualty due to fire mistakenly directed at U.S. or allied troops; aka Friendly Fire

Mission: A specific assignment or task that a person or element is sent to accomplish

Mister: The correct prefix (followed by their name) when addressing a Warrant Officer, "Sir" is not used

Mity Mite: Agricultural backpack-sprayer used for smoke or tear gas to flush enemy from tunnels; not Mighty Mite

Mk 4: Folding-Fin Aerial Rocket (FFAR); 2.75-inch, 4ft long, 17lb, pod-fired, unguided, air-to-ground missile

MK1956: Pistol Belt on which one hangs web gear, ammo pouches, a First Aid kit, canteens, etc.

MK2: A U.S. 21oz, pineapple-shaped, cast iron, fragmentation hand grenade used early in the war: aka Pineapple

MK3A2: See Concussion Grenade

MMAF: Marble Mountain Air Facility; MAG-16 airbase

MOGAS: Motor Gas; gasoline

MOH: *Medal of Honor*; America's highest military medal for heroism; aka CMH

Moi: Vietnamese slang for savage; derogatory term referring to Montagnards

Mojave: See CH-37

Monday Pills: Once-a-week, anti-malarial pills; aka Dapzone and Horse Pill

Monitor: Several variants of converted LCM-6, heavily armored & gunned, single-turret, 60', shallow-draft gunboat

Monopoly Money: See MPC

Monsoon: The Vietnam rainy season

Montagnard Bar: Facetious term for a Tropical Hersey Bar found in some sundry packs; aka Doggie Chocolate

Montagnard: A member of one of 54 indigenous hill tribes of Southeast Asia

Montagnards: French term meaning *People of the Mountain*

Moon Floss: Slang term for toilet paper

Moon: To show your gluteus maximus in public either as a prank or to show contempt

Moose: Disrespectful term for a Vietnamese mistress

MOOSE: Move Out Of Saigon Expeditiously; the relocating of 1,000's of troops from Saigon to basecamps in 1967

Mortar: A base-plated, muzzle-loaded, tubed, high-angle firing, short-range, crew-served gun

MOS: Military Occupational Specialty; a code-numbered job title

Most Rikki-Tik: To act very quickly; see also Rikki-Tik

Mother Green: An affectionate term for the Marine Corps

Motivation Platoon: Punishing USMC boot camp special element for transforming Shitbirds into disciplined recruits

Move Out: To begin a patrol or military operation

MP: Military Police

MPC: Military Payment Certificate; a scrip used instead of U.S dollars; see also Monopoly Money or Funny Money

Mr. Charles: Slang for a very good enemy fighter; see also Chuck and Charlie

MRF: Mobile Riverine Force; joint Army-Navy, all-weather, strike-missioned fleet; aka Brown Water Navy

MSB: Mine Sweeper, Boat; Brown Water Navy mine sweeper

MSO: Minesweeper, Ocean *Avenger-Class*; had a fiberglass-covered wooden hull to avoid magnetic mines

MSR: Main Supply Route; the track or series of roads and trails over which troops/supplies enter operational areas

MULE: Multi-Utility, Light Equipment; Willys M274, low, half-ton, 4WD, highly-versatile, platform-like vehicle

MUST: Medical Unit, Self-contained, Transportable; an truck- and air-moveable, inflatable, 40-bed surgical hospital

Mustang: An enlisted-grade man who gets commissioned to officer-grade

Muster: Roll Call formation; gather up or meet or meeting

MUTT: M151 Military Utility Truck; ¼-ton, 2,400lb, 4WD, utility, unibody, personnel transport truck; aka Jeep

Muzzle Flash: Hi-pressure, hi-temperature, and flash following a projectile from the forward end of a gun or rifle

MXX991/U: Right-angled flashlight with a 3-way switch (Off-Blink-On) and 3 lens filters

N

Nail Round: An aerial rocket or artillery round filled with flechettes; see also Beehive and Satan's Toothpicks

Nail: Lingo for kill or destroy; see also Blow Away or Grease or Hose or Waste or Zap

Nails: Flechettes or Satan's Toothpicks

Nam, The / Nam: Clipped term meaning Vietnam

Napalm: Highly-volatile, sticky, soap-gelled gasoline, incendiary weapon used in flamethrowers and bombs

Nape: Slang for Napalm

NAS: Naval Air Station

National Liberation Front: See NLF

Native Sport: Satirical term for the hunting of Viet Cong

NATO: North Atlantic Treaty Organization; U.S., Canada, and various European country alliance for joint security

NAV: Navigation; navigator

NAVSUPDEP: Naval Supply Depot; aka NSD

NBC: Casualty code for *Non-Battle Casualty*

NBC: Nuclear, Biological, Chemical; refers to weapons of this mass casualty causing ilk

NCO: Non-Commissioned Officer; enlisted rank of E-4 and E-5; see addendums

NCOIC: Non-Commissioned Officer In Charge

NDP: See Night Defensive Position

Newbie / Newby: Anyone who is fresh to the war or to a unit; an FNG or a Noob

NGFS: Naval Gun Fire Support; shore-bombardment using naval ship-launched artillery or missiles

NGO: Non-Governmental Organization

Night Defensive Position: Hastily dug series of perimeter holes for a unit to overnight in while in the boonies

NJP: Non-Judicial Punishment; a commander's way of dealing with minor crimes without a formal Court Martial

NLF: National Front for the Liberation of the South; the Viet Cong; term distinguishes the NLF from the PAVN

NLT: No Later Than

No Joy: Radio code meaning a pilot cannot see an object, or he cannot hear a radio transmission

No Man's Land: The area located between two antagonists in a battle

No Sweat: Response meaning *I can do that easily*; something done with little difficulty

NOK: Next Of Kin

Nomex: Flame-resistant uniform material worn by flight crews

Non-Hacker: An individual who is incapable of completing a task or challenge usually physical or stressful in nature

Non-Judicial Punishment: See NJP and Article 15 and Office Hours and Mast and Captain's Mast; see addendums

Non-Qual: A person who fails to pass a course or test and is therefore ineligible for certification

Noob / Noobie: A person new to Vietnam or new to the unit or job; an FNG

NOR: Not Operationally Ready

North Atlantic Treaty Organization: See NATO

North, The: Lingo for NVN

NSA: Naval Support Activity; unit responsible for such operations as hospitals, fuel storage, or port facilities

NSC Working Group On Vietnam: Explored objective decision options for LBJ based on statistical probabilities

NSC: National Security Council

NSD: Naval Supply Depot, aka NAVSUPDEP

Nuclear Triad: Cold War term for U.S.'s strategic bomber aircraft, ICBM's, and submarine-launched ballistic missiles

Number One: The best

Number Ten: The worst

Number Ten-thousand: Absolute worst of the worst; aka *Number Ten-thou*

Nung: One of 54 tribes of Montagnard (mountain people) found in Vietnam, the Nung were of Chinese descent

Nuoc Mam: A Vietnamese fermented fish sauce; aka Armpit Sauce, Nước Mắm

Nut To Butt: Line up tight with one's testicles close to the ass of the person before them; see Asshole to Bellybutton

NVA: North Vietnamese Army; lingo for a single communist North Vietnamese soldier

NVN: North Vietnam

O

O-1: Cessna *Bird Dog*; single-propeller, single-engine, tandem-seat, light observation, FAC aircraft

O-2: Cessna *Skymaster*, twin-engine, two fuselage-mounted props (1 fore, one aft), tandem-seat, light FAC aircraft

O-3: Short for a Marine rifleman; refers to the "zero-3" in a Marine's MOS of 0311; pronounced *Oh-3*

OAS: Organization of American States; formed in the Cold War as a buttress against the spread of communism

Obliteration Bombing: See Carpet Bombing

Observation Post: See OP

OCO: Office of Civilian Operations: Embassy-controlled authority over all civilian agencies operating In-Country

OCS: Officer Candidate School (Marine Corps)

OD: Officer of the Day

OD: Olive Drab; standard military green color

Off The Wall: Slang for strange or unusual or weird

Office Hours: Non-Judicial Punishment for minor offenses with no criminal proceeding required

Office Pogue: A person who works behind a desk

Officer Material: A sardonic term for a useless enlisted man

Officer Of the Deck: Current watch officer in charge of a ship; aka OOD

OH/EX: On Hand/Expended; inventory classification code to compare ammunition amounts

OH: Observation Helicopter

OH-58: Bell *Kiowa*; a light observation and target acquisition and attack helicopter

OH-6A: Hughes *Cayuse*; a nimble, single-rotor, light observation and target acquisition helicopter; see Loach

OIC: Officer In Charge

OJT: On the Job Training

Old Boots: Non-derogatory term for personnel serving In-Country for a long time

Old Lady: A slang term for a wife or girlfriend

Old Man: Slang for Commanding Officer of a unit; aka CO

Old Salt: An experienced military man

Old Timers: Non-derogatory term for personnel serving In-Country or in the military for a long time

Olive Drab: See OD

On Line: When individuals or elements are abreast

Ontos: Allis Chalmers M50A1 light, nimble, tracked, anti-tank vehicle with six 106mm recoilless rifles, 1,500yd range

OOA: On Or About

OOD: See Officer Of the Deck

OP: Observation Post; daytime hole or hide for 2 or 3 troops outside their unit's perimeter to act as early warning

OPCON: Operational Control

OPCON'ed: Placed under the Operational Control of another unit

Operation Baby Lift: 1975 airlift out of Vietnam of some 2,000 orphans/children due to the looming fall of Saigon

Operation Barrel Roll: Code name for the long-term bombing campaign against enemy targets in Laos

Operation Enhance: The 1972 massive U.S. delivery of war materials to SVN prior to the Paris Peace Accords

Operation Homecoming: Code name for the plan to repatriate American POW's after the Paris Peace Accords

Operation Menu: Code name for the bombing campaigns targeting enemy base camps in Cambodia; see Menu

OPLAN 34-A: Secret covert MACV-SOG and CIA warfare action program against the DRV

OPORD: Operational Order; formal directive explaining the mission, what the unit will face, support assets, etc.

Option IV: Top secret plan to evacuate Saigon by helicopter

OR: Operating Room

Order Of Battle: The various elements of an army and their arrangement and disposition for battle

Ordnance: Any explosive, including ammunition, bombs, napalm, grenades, rockets, flares, etc.

ORLL: Operations Report, Lessons Learned; a unit's after-action commentary or routine quarterly account

OSD: Office of the Secretary of Defense

OTHD: Other Than Honorable Discharge; see addendums

Other Than Honorable Discharge: See addendums

OTS: Officer Training School (Army)

Otter: See DHC-3

Out Of Sight: Slang term meaning for exceptional, remarkable, or absolutely amazing; aka Outta Sight

Out Sheet: See DD-214

Out-Country: Still in Southeast Asia, but not in Vietnam; see also In-Country

Outgoing: Artillery, mortar, or rocket fire launched against the enemy

Outstanding: Terrific or remarkable or well done

Outta Sight: Slang term meaning for exceptional, remarkable, or absolutely amazing; aka Out Of Sight

OV-10: See Bronco

Over The Fence: Lingo for past the border from SVN into NVN or into Laos or into Cambodia; aka Across The Fence

Over The Hill: See UA and AWOL and Desertion

Over The Hump: Lingo for more than halfway through something

Over The Pond: Lingo for crossing the Pacific Ocean on the way to Vietnam

Overhead: Navy lingo for the ceiling

P

P: See Piaster

P-38: Tiny, folding C-Ration can opener; aka John Wayne Can Opener

PAB: Plastic Assault Boat; fiberglass 16ft boat

PACEX: Pacific Exchange; a mail-order company similar to a Sears or Montgomery Ward

Pacification: A program to gain support of villagers for the GVN and destroy the influence of the VC and NVN

Paddy Foot: Caused by prolonged wetness resulting in open ulcers, fungal infection, etc.; see also Immersion Foot

Paddy: A field, diked and flooded, for growing rice

Padre: Affectionate term for a military chaplain; literally *Father*

Papasan: Bastardized Japanese word meaning honorable father; an older Vietnamese man

Parade Deck: See Parade Ground; aka Grinder

Parade Ground: Flat expanse for marching, drilling, and for ceremonies; aka Grinder and Parade Deck

Paris Peace Accords: The 27 January 1973 agreement to end fighting in Vietnam

Passageway: Navy lingo for a corridor or hallway

Pathet Lao: Laotian communist forces; aka PL

Pathfinders: Specially-trained teams inserted into remote sites to facilitate upcoming airborne operations

Patrol Boat, River: See PBR

Patton: M48, 45-ton, 30mph, armed with a 90 or 105mm gun, 50-ton, main battle tank

PAVN: People's Army of (North) Vietnam; the NVA; term distinguishes the NVA from the NLF

PAX: Passenger

Payback: A wrathful term meaning retaliation or retribution or revenge or vendetta or vengeance

Payback's A Motherfucker: Objective statement after retribution has been carried out

PBR: Pabst Blue Ribbon beer

PBR: Patrol Boat, River; 28.5kph, 31ft, twin-diesel, waterjet-driven, shallow-draft, fiberglass, combat watercraft

PCF: Patrol Craft, Fast; see Swift Boat

PCS: Permanent Change of duty Station

PDO: Priority Dust Off; serious injury, but not an emergency

PDQ: Pretty Damn Quick

Pecker Check: Slang term meaning an examination for venereal disease; aka Short Arms Inspection

Pecker Checker: Slang term for the medical person who examines a penis for venereal disease

Peckerwood: A derogatory racial reference used by Blacks to denote a White person; aka Chuckerwood

Pedi-Cab: A bicycle-powered rickshaw; see Cyclo

Pedro: Kaman *Huskie* HH-43F twin 2-blade contra-rotor, single turboshaft engine, 120mph, AF rescue helicopter

Pentagon Far East: Nickname for MACV HQ in Saigon; aka Disneyland Far East

Pentagon: Department of Defense HQ; so-called because of the building's 5-sided shape; aka Disneyland East

People Sniffer: Highly-sensitive, urine/sweat sensor transmitter devices employed to detect enemy in remote areas

People's Revolutionary Party: See PRP

Perimeter: A position's, or base's, outer boundary

PF: See Popular Forces

PFT: Physical Fitness Test

PGM-5: A Brown Water Navy boat with a turret-mounted flamethrower; effective range of 300-yards; see Zippo

PH: See Purple Heart

Phantom: See F-4

Phase Line: Trace, such as a road, stream, or map feature that is easily defined for use as a control point in battle

Phoenix Program: Assassination, subversion, and terror stratagem to eliminate NLF operatives

Phonic Alphabet: See addendum

Phrog: Affectionate nickname for a CH-46 helicopter

Piaster: Basic monetary unit (P) of Vietnam during French colonization, slowly phased out in the 1960's

Pick Up Zone: An LZ; aka PZ

Pickle: Slang for the act of depressing the *Release* button in a cockpit thereby dropping bombs from an aircraft

Piece: A gun, rifle, knife, or other type of weapon

Pierce Arrow: Code for the retaliatory air attacks on NVN for the 1964 attacks on U.S. ships in the Tonkin Gulf

Pig: M60; 7.62mm, belt fed, air-cooled, 600 rounds per minute, light machinegun; see also Hog

Pineapple: See MK2

Pink Team: Irregular warfare, wherein a specially-trained, hunter-killer team exercises independent maneuver

Piss & Punk: Navy lingo for water and bread rations given to a prisoner in solitary confinement

Piss Cutter: Long, tent-shaped garrison hat

Piss Tube: A urinal made from a pipe sticking out of the ground into which one urinates

Pissed / Pissed Off: Slang for quite angry

Pits, The: Slang term for *worst ever* or for something depressing

PL: See Pathet Lao

PLAF: People's Liberation Armed Forces; the totality of the Viet Cong infrastructure, independent of the PAVN

Plain of Reeds: A 2,500sqmi inland Mekong Delta swamp area and often-contested Viet Cong sanctuary

Plastic: Lingo for a soft, malleable, chemical explosive; see C-4

Plenty Cheap Charlie: Vietnamese use of bastardized English for a skinflint; see also Cheap Charlie

PM-41: Soviet 82-PM-41, 82mm smooth-bore, muzzle-loading, indirect-fire, 3.1mi range, medium-weight mortar

PM-43: Soviet 120-PM-43, 120mm smooth-bore, muzzle-loading, indirect-fire, 3.5mi range, heavy-weight mortar

Pneumatic Fluid: A fruitless-search prank on FNG's sent to find this non-existent item

Pocket Money: Code name for the mining of Haiphong and other harbors & ports in May 1972; see also Duck Hook

Pogey Bait: Junk food; candy or other sweets

Pogue: Disparaging term for an office worker; a languid individual

Point: The forwardmost man leading a column of infantrymen on a combat mission; aka Pointman

Pointman: See Point

POL: Petroleum, Oil, and Lubricants

Police / Police Up: To clean or de-litter an area

Police Call: The order to clean up or de-litter an area

Poncho Liner: Quilted, rip-stop fabric, lightweight, quick-drying insert for a poncho; most often used as a blanket

Poncho Litter: Makeshift Stretcher on which to drag or carry the wounded or dead

Poncho: Hooded, outer garment of nylon or other synthetic to stay warm and dry or to use as a makeshift shelter

PONR: Point Of No Return; refers to the doomed moment when an aircraft no longer has enough fuel to RTB

Poop: Intelligence or information; see also Dope and Intel and Scoop and Skinny and Word

Poor Man's Artillery: Lingo for a mortar

Pop A Smoke: The act of igniting a smoke grenade to identify an LZ, friendly position, target, or wind direction

Pop Flare: A handheld aerial illumination flare, launched by whacking the bottom of the device with your palm

Pop: Ignite, trigger, or fire a smoke grenade or flare

Popeye: Radio code from an aircraft which proclaims that they are flying by instruments only; see IFR

Popular Forces: Self-defense corps of part time local militia that protected SVN hamlets and villages; see also PF

Pop-Ups: A pop-up flare; see Pop Flare

Port Hole: Navy lingo for window, does not refer to Port

Port: Navy term for the left when facing forward

Pos: Short for Position

Position: Location

Pot Head: Slang for an individual who uses marijuana; aka Head

POW: Prisoner Of War

Prairie Fire: Code phrase indicating a SF or SOG group is compromised and evading, or about to be overrun

PRC-25: Portable Radio, Communications-25; the standard portable, back-carried, infantry FM radio; aka The Prick

PRC-77: Portable Radio, Communications-77; portable, back-carried, VHF/FM radio, encryption-capable

Prep: Preliminary bombardment by air, artillery, mortar, or naval gunfire of a landing beach or landing zone

Prick, The: See PRC-25

Prickly Heat: See Miliaria

Primo: The best; top of the line

Profile: Medical term temporarily relieving a person from performing certain military tasks due to injury or illness

Project 100,000: Ruse to recruit those known to be below military standards to fill escalating war manpower needs

Prop Wash: A fruitless-search prank on FNG's sent to find this non-existent cleaning solvent

Prop Wash: The forceful airstream generated by an aircraft's turning rotors or propellers; aka Rotor Wash

Protective Reaction Strike: Those air attacks below the 20th Parallel in NVN that protected reconnaissance missions

Provider: See C-123

Province: U.S. state equivalent in Vietnam

PRP: People's Revolutionary Party; provided leadership for the newly formed VC in SVN, operated under COSVN

PSDF: People's Self Defense Force; see PF

PSP: Perforated Steel Planking; 15" x 10' interlocking (hook and slot), staggered mats for rapid laying of runways

Psychedelic: A bright rainbow of colors and patterns associated with Flower Children and Hippies

Psy-Ops: Psychological Operations; see Chieu Hoi and Hearts & Minds and Phoenix

PT: Physical Training; any exercise

PT-76: Soviet, 14.6-ton, 76.2mm gun, 27mph, 6mph in water (hydrojet), 230mi range, light amphibious tank

Pucker Factor: Facetious measure of anal sphincter tightness correlated to the degree of danger encountered

Puddle Pirate: Erroneous term for a Coast Guardman, because they supposedly never operate in deep water

PUFF: Call sign for a Douglas AC-47 *Puff, the Magic Dragon*; gunship rigged with 3 mini-guns and flares; aka Spooky

Pugil Stick: Padded-on-both-ends rod, rifle-length, to learn bayonet/stock combat by two men smashing each other

Puke: Demeaning term for a depraved, sorry excuse of a human being

Puking Buzzard: Derisive term describing the Screaming Eagle shoulder patch design for the 101st Airborne

Punji Pit: A fall-into boobytrap consisting of sharpened wood or metal spikes pointed upward in a camouflaged hole

Punji Stake: Sharpened wood/bamboo used in some boobytraps, often set where a person may likely dive for cover

Purple Heart: Military medal awarded to those wounded or killed in combat

PX: Post Exchange; a store on a Navy or Marine Corps installation; see also BX

PZ: A helicopter Pickup Zone; an LZ

Q

QRF: See Quick Reaction Force

Quad-50: M55 turret-mounted, solenoid-fired, four .50cal Browning machinegun platform; aka Meat Grinder

Quarters: Military living space such as barracks, squad bays, etc.

Queer: Derogatory term for homosexual

Quick Reaction Force: Infantry unit ready to respond within minutes (usually by helicopter) to an enemy attack

Quick Reactionary Force: The 2nd squad in a file; would deploy on-line to assist when the lead element was hit

Quonset Hut: Curved steel buildings 20' x 48' used for barracks or storage, first built in Quonset Point, RI

R

R & R: Rest & Recreation; a 5- or 7-day, once-per-tour vacation to one of ten countries; see also In-Country R&R

RA: Regular Army enlistee (not a Draftee), noted by the prefix *RA* on one's service number; see also US

RAC: Radio Alphabet Code; the Military Phonic Alphabet; see addendums

Rack: A Berth, Bunk, or Cot; a bed

Rain Locker: Navy lingo for a shower room

Rallier: An enemy deserter who surrendered to SVN under the Chieu Hoi Program

Ranch Hand: Code name for the U.S. defoliation campaign

Ranger: Commandos specially-trained for insertions, pathfinding, repelling, and LRRP

Rank: A side-by-side row of individuals or elements

Rank: Your grade or level in the Chain of Command, such as Sergeant or Major; see addendums

RAP: Rocket Assisted Projectile; uses a rocket motor as independent propulsion thereby extending its range

Rap: Slang for an affable, meaningful conversation

Rapid Reaction Force: See RRF

Rappel: A descent, usually from a helicopter, by rope

Razor Grass: Sharp-bladed, concertina-edged grass that would cut uniforms and tear skin

Razor Wire: Concertina wire with pointed razors designed as an obstacle that cuts and entangles enemy troops

RBF: See Reconnaissance By Fire

RC-292: Radio Component 292; portable, 41½ft, ground plane antenna used to extend the range of field radios

RCA: Riot Control Agent; see CS

Ready Reaction Force: (RRF) lightly-armed platoon, pre-placed, usually airlifted, to quickly respond to emergencies

Real Life: Slang for being a civilian; living not in the military

Rear, The: Usually a large base, outside the immediate danger of the bush

Recon By Fire: See Reconnaissance By Fire

Recon: An individual Reconnaissance Marine; see also Battalion Recon and Force Recon

Recon: See Reconnaissance

Recondo: Literally *Reconnaissance, Commando, and Doughboy*; U.S. Army Ranger/LRRP school

Reconnaissance By Fire: The firing at suspected enemy positions in hopes of locating them

Reconnaissance In Force: Larger than normal scouting patrol to find and test the enemy's strength and assets

Reconnaissance: The act of covert observation of enemy activity and the gathering of Intelligence without Contact

Recruit Division Commander: Chief Petty Officer and Petty Officer (1st or 2nd Class) Navy recruit trainer

Red Alert: The last and most urgent of all alerts, signaling that a known enemy is about to attack

Red Ball: Lingo for a hard-packed (well-travelled) jungle trail indicating substantial enemy use; see also Redball

Red Beach II: Site on Da Nang Bay where on 8 March 1965, Marines made the first amphibious landing of the war

RED HORSE: Rapid Engineering Deployment & Heavy Operational Repair Squadron, Engineering; construction team

Red Leg: Artilleryman; leftover U.S. Civil War term referring to the red piping on a Union artilleryman's trousers

Red Tracer: A visible color streak left momentarily in the air to track the path of a bullet (NATO); see Tracer

Redball: Lingo meaning to quickly deliver critical supplies; see also Red Ball

Reefer: Navy term for a large refrigerator or a refrigerated room; slang for marijuana

Reenlist: To sign on again for a two-, three-, or four-year period of active duty

REFRAD: Released From Active Duty

Regimental Landing Team: See RLT

Regular Army: Those soldiers who enlisted as opposed to those who were conscripted; see RA

Rein: Short for Reinforced

Reinforced: Usually when a unit is temporarily strengthened with additional material, equipment, or personnel

REMF: Rear Echelon Motherfucker, derogatory name for a person living in the Rear; see In The Rear With The Beer

Remington Raider: Derisive term for a typewriter-manned REMF; aka Smith-Corona Commando

Repeat: Radio code word requesting another salvo of artillery rounds at the same target location

Repo Depot: Lingo for the staging area where new replacements arrive and are processed

Repose, USS: One of two Navy hospital ships that plied the oceanic waters off Vietnam; see also Sanctuary, USS

ReQual: Mandatory, yearly, rifle requalification for every Marine (waived while in Vietnam)

Request Mast: Procedure in which an individual can have his serious problem heard up the Chain of Command

Reserve Officer Training Corps: See ROTC

Resupp: Resupply

Re-Up Bird: A jungle bird (Blue Eared Barbet) with a call that sounds like *Reeeee-Up!* (which means to reenlist)

Re-Up: See Reenlist

Reveille: Sunrise bugle signal to awaken military personnel, usually to gather in formation for morning roll call

Revetment: A three-sided earth or built up barrier used as blast protection, usually for parked aircraft

RF/PF: See Ruff-Puff

RF: See Rural Forces

RF-101C: McDonnell *Voodoo*; supersonic, single-seat, twin-turbojet, unarmed, photo reconnaissance aircraft

Rhade: One of 54 tribes of Montagnard (mountain people) found in Vietnam

Rice Paddy Daddy: An offer of what one could be if a pretty woman was to take a liking to him, a sugar daddy

Rice Wine: A cheap alcoholic drink made from rice as its base ingredient

Ricky Recon: A grunt teasing term for Reconnaissance Marines derived from Ricky Ricardo on the *I love Lucy* show

Ride: Slang for a personally-owned mode of transport such as a car, motorcycle, or truck

RIF: See Reconnaissance In Force

Rifleman: The Marine infantryman; every Marine's primary MOS regardless of their subsequent MOS certification

Right On: A slang term meaning correct or exactly or righteous or agreement

Righteous: Slang for a morally correct or virtuous act; excellent

Rikki-Tik: Slang for soonest; to not delay; to act very quickly one would say Most Rikki-Tik

Riot Gas: See CS

Rip Off: Slang for a scam, a theft, or to steal from

River Patrol Boat: See PBR

River Rats: Affectionate slang for Brown Water Navy forces

Riverine Force: See MRF

RLG: Royal Lao Government

RLO: Real Live Officer; derisive term used by helicopter Warrant Officers for a commissioned-officer pilot

RLT: Regimental Landing Team; USMC infantry task force ready for amphibious/helicopter insertion into combat

Road Guards: Intersection safety wherein 1 or 2 trainees break from a marching/running formation to block traffic

Road Sweep: A comprehensive check (usually at first light) of thoroughfares for mines and the elimination thereof

Roadrunner: A Marine Corps mine-sweeping team

Rock And Roll: Lingo for stepping off on a mission

Rock And Roll: Lingo meaning to fire a weapon in full-automatic mode

Rock Ape: Legendary rock-throwing primate; probably a Tonkin snub-nosed monkey

Rocket Rainmaker: A 207' ship with equivalent firepower of five destroyers; see also LSMR

ROE: Rules Of Engagement

Rog: Short for Roger

Roger That: Same as Roger; loose slang for *I agree with you* or *Right On*

Roger: Radio code term for *I understand*

ROK Marine: A Republic of (South) Korea Marine

ROK: Republic of (South) Korea

ROK's: Lingo for Republic of Korea allied forces

Roll Call: A gathering into troop formation for enumeration; a calling out of names to account for all persons

Rolling Thunder: Code name for the aerial bombardment of targets in NVN

Rome Plow: A D7E Caterpillar 26-ton bulldozer fitted with a huge sharp blade used to clear very heavy vegetation

Romp `N Stomp: Army derisive term for marching or drill field maneuvers

RON: Remain Over Night

Rotate: A term used to denote the return to the World at the end of one's tour of duty

ROTC: Reserve Officer Training Corps; prep program for college students to become officers; see also JROTC

Rotor Heads: Condescending term for personnel assigned to helicopter crews

Rotor Wash: A fruitless-search prank on FNG's sent to find this non-existent cleaning solvent

Rotor Wash: The forceful airstream generated by a helicopter's turning blades; aka Prop Wash

Rottencrotch, Rosie or Suzy: Girlfriend of Joe Shit the Ragman

Rough Rider Convoy: Armed procession of military vehicles moving through dangerous territory as fast as possible

Rough Rider: Individual assigned to perilous convoy duty

Round Eye: Slang term for any female without an oblique eye shape; generally referring to non-Asian females

Round(s) On the Way: Urgent radio alert that artillery, mortar, or naval gun shells have been fired as requested

Round: An artillery or mortar or naval gun munition or bullet, whether assembled or its projectile; see also Shell

Route Step: Not a run, but a very fast march without cadence or tempo

RPD: Soviet *Ruchnoy Pulemyot Degtyaryova (RPD)* Soviet, 700 rounds-per-minute, belt-fed, light machinegun

RPG Screen: Chain-link fencing erected for protection from grenades

RPG: Rocket-Propelled Grenade; fired from a RPG-2 or RPG-7

RPG-2: Soviet anti-tank, rocket-propelled, shoulder-fired, 70mm warhead, 40mm grenade launcher; aka B40

RPG-7: Soviet anti-tank, rocket-propelled, shoulder-fired, 85mm warhead, 40mm grenade launcher; aka B41

RPK: Soviet *Mikhail Kalishkinov* Soviet, 600 rounds-per-minute, magazine-fed, light machinegun; replaced the RPD

RRF: See Ready Reaction Force

RSSZ: Rung Sat Special Zone; 485sqmi mangrove swamp just south of Saigon, often-contested, Viet Cong sanctuary

RT: Recon Team

RTB: Return To Base

RTO, The: The Radioman

RTO: Radio-Telephone Operator; generally a field unit's radioman who carried the PRC-25

Rubber Bitch / Rubber Lady: Satirical nickname for the UVG-0029 air mattress

Ruck: See Rucksack

Rucksack: Standard issue infantry backpack; tubular aluminum frame with 3 or 4 water-resistant pockets for gear

Ruff-Puff: Condescending term for the poorly trained and equipped Rural Forces and Popular Forces; see RF and PF

Rules of Engagement: Edict on circumstances/limitations/implications of when, where, and how combat can occur

Rumor Control: Cold War plan for identifying, and mitigating, disruptive gossip before it causes harmful escalation

Rung Sat Special Zone: Rừng Sát; see RSSZ

Runner: Lingo for a messenger

Rural Forces: SVN militia-type groups that guarded critical points such as bridges and ferries; aka RF

RVN: Republic of (South) Vietnam

RVNAF: Republic of (South) Vietnam Armed Forces

S

SA: See Small Arms

SA-2: A 4,850lb, 3.5Mach, 60,000ft ceiling, Soviet surface-to-air missile (SAM); also its launcher

Sabot: Device to keep a smaller projectile (usually an armor-piercing round) centered in a larger-bored gun

SAC: Strategic Air Command; the Cold War command for the Nuclear Triad and strategic air reconnaissance

SAC: The Command controlling B-52 saturation (carpet) bombing missions over both Vietnams, Laos, and Cambodia

Saddle Up: A term meaning put on one's combat gear and be ready to move out

SAF: Small Arms Fire

Saigon Tea: Expensive shot glassfuls of colored water purchased for a Vietnamese female 'companion' in a bar

Saigon Warrior: A soldier who never participated in combat

Saigon: The capital city of South Vietnam

Sailor: A catch-all term for anyone in the Navy

Salad Bar: The patch of military ribbons on a military dress uniform

Salty: Lingo for generally profane, opinionated smart-aleck wordage; an adjective for the person using such words

Salvo: A simultaneous firing of battery guns aimed at the same target; aka Volley

SAM Song: Audible warning that a surface-to-air missile has been fired and is tracking your aircraft; see MAWS

SAM: Surface-to-Air Missile

SAM-7: Soviet, portable, lightweight, shoulder-fired SAM that debuted late in the war

Same-Same: Slang term meaning alike or equivalent or identical

Sampan: A general term for a small, flat-bottomed Vietnamese boat of varying designs

Sanctuary, USS: One of two Navy hospital ships that plied the oceanic waters off Vietnam; see also Repose, USS

Sandbagging: See Lollygagging

Sand Table: Mission-planning model built to topographical scale of an operational area

Sapper: Enemy commando/infiltrator trained in demolitions

SAR: Search & Rescue; in Vietnam it is usually CSAR; aka Combat Search & Rescue

Sát Còng: Vietnamese phrase meaning *Kill Communists*

Satan's Toothpicks: See Flechettes

Satchel Charge: High explosive pack thrown or dropped by an enemy Sapper

Saturation Bombing: See Carpet Bombing

Saturation Fire: Massive bombardment or gunfire to eliminate enemy activity or installations

Say Again: Radio code for *What?* or a request to restate what was just said

Scare America: Slang for Air America

Scoop: Intelligence or information; see also Dope and Intel and Poop and Skinny and Word

Score: To get something needed, wanted, or of value

Scorpion: See M56

Scout Dog: Specially-trained dogs used to detect or sniff out the enemy

Screaming Eagle Replacement Training School: See SERTS

Screaming Eagle: A member of the 101st Airborne Division

Screaming Shits: Diarrhea so urgent that one cannot control the timing or power of erupting hot, liquid fecal matter

Scrounge: Lingo for unofficial pilfering or bartering for needed unavailable items; see also Cumshaw

Scullery: A place in a mess hall where utensils, dishes, and trays are washed

Scuttlebutt: Navy lingo for a drinking fountain or rumors or gossip

Sea Dragon: Navy bombardment of land targets in NVN and the interdiction of men and supplies infiltrating by sea

Sea Duty: Being assigned to a ship, as opposed to land-based duty

Sea Knight: See CH-46

Sea Lawyer: A self-aggrandizing, self-appointed expert on a subject or collection of subjects

Sea Stallion: See CH-53

Sea Story: A conjured tale of great embellishment

SEA: Southeast Asia

Seabag: A large oblong, canvas bag to store and carry all of a Marine/Sailor's uniforms; see also Duffle/Dufflebag

SeaBees: Navy Construction Battalion; aka CB's

Sea-Going Bellhop: Sardonic term for a Marine assigned to the bridge area of a ship

Seahorse: See CH-34 and HUS-1

SEAhut: Southeast Asia Hut; plywood and tin, one-story housing for rear-area troops

SEAL Team Assault Boat: Usually a small, fast, 10-men-or-less, insertion watercraft; aka STAB

SEAL: The Navy's Sea-Air-Land special warfare maritime covert commandos and advisors to ARVN special forces

Search & Clear: A more politically-correct and palpable term for Search & Destroy

Search & Destroy: Inserting troops into hostile areas, finding and killing the enemy, and leaving soon thereafter

Search & Rescue: See SAR

Search And Avoid: Sarcastic phrase referring to the efficacy of ARVN combat missions

SEATO: Southeast Asia Treaty Organization; operated cultural and educational programs

SECAF / SAF: Secretary of the Air Force

SECARMY / SA: Secretary of the Army

SECDEF / SecDef: Secretary of Defense

SECNAV: Secretary of the Navy

SECSTATE: Secretary of State

Secure: To tie down or lock up; to safeguard or make safe an area or structure or thing; cease; shut

Sedang: One of 54 tribes of Montagnard (mountain people) found in Vietnam

SEL: Suspected Enemy Location

Selective Service System: Agency responsible for military manpower conscription; see also Draft, The

Semi-Automatic: General reference to any gun or rifle that will fire once with each trigger pull without reloading

Semper Fi: Shortened term for Semper Fidelis

Semper Fidelis: Latin for *Always Faithful*, the Marine Corps motto

Semper Gumby: Slang Marine term meaning *Always Flexible*

Semper Scrotus: Slang Marine term meaning *Always On the Ball*

Sentry: Guard who controls access to an area via gates or other openings in a perimeter and observes/hears all

SERTS: In-Country five-day training and orientation school for all troops new to the 101st Airborne Division

Serviceman's Readjustment Act: See GI Bill

Seven-Eighty-Two Gear: See 782 gear

Seventeenth Parallel: See 17th Parallel

SF: Special Forces

SFD: Surprise Firing Device; a boobytrap

SFOB: Special Forces Forward Operating Base

Shackle: Encryption; an alphanumeric code system used on radios to disguise sensitive information

Shades: Slang for sunglasses

Shadow: See AC-119; aka Flying Boxcar

Shafted: Slang term meaning cheated, screwed, or otherwise treated unfairly

Shake `N Bake: High Explosive rounds and Napalm cannisters impacting in the same attack

Shake `N Bake: Soldiers who attended NCO academies for a higher rank, usually only those in the combat arms

Shake `N Bake: A soldier fast-tracked for promotion not on merit, but to fill leadership vacancies

Shake `N Bake: Regular soldier who went thru Recondo School In-Country, but not full-blown training at Ft. Benning

Sham Timing: Slang for the act of taking extended time for errands/details that should take much less time

Shamming: Slang for pretending, faking, or goofing off

Shaped Charge: A one-direction or fan-formed explosive

Shavetail: Slang for new 2nd LT's; refers to the act of shaving the tails of newly-broken, thus unproven, Army mules

Shawnee: See CH-21

Shell: An artillery or mortar or naval gun munition whether assembled or its projectile; see also Round

Shelter Half: One-half of the canvas of a 2-man tent with 5 tent pegs, 1 guy line, and one 3-section pole

Shelter Tent: A 2-man tent consisting of 2 shelter halves, 10 tent pegs, 2 guy lines, and two 3-section poles

Sheridan: GMC *Sheridan* M551, 152mm, 17 ton, 43mph, aluminum, swimmable, light tank

Shining Brass: Cross-border ops into Laos by U.S./SVN Special Forces to observe and disrupt NVA entry into SVN

Ship Over: Navy lingo to Reenlist; aka Re-Up

Shit Can: Slang for a Garbage Can

Shit Can: Slang meaning to throw away or otherwise dispose of

Shit On a Shingle: Lingo term for chipped beef in heavy cream gravy on toast; aka SOS

Shit Sandwich: Slang for some form of violent activity such as combat or major chaotic trouble; aka Shit Storm

Shit Storm: Slang for some form of violent activity such as combat; sometimes a temper tantrum; aka Shit Sandwich

Shit, Shower, and Shave: See Triple S

Shitbird: Slang for an individual who is an undisciplined, non-motivated, worthless, waste of human existence

Shithook: Army lingo for a Chinook; aka CH-47

Shitter: An outhouse, usually found in rear areas (no pun intended) built over 55-gal half barrels for collection

Shoot `N Scoot: Scornful term for the enemy tactic of firing and then moving quickly away to evade counterfire

Shoot The Breeze: Slang for to converse casually

Shoot The Shit: Slang for to converse casually

Short Arms Inspection: Slang for an examination for venereal disease; aka Pecker Check

Short Blanket Roll: One blanket, one guy line, 5 tent pins, 1 tent pole, 1 shelter half, and 2 blanket-roll straps

Short Orbit: Lingo for a tight circular path for an aircraft overhead

Short Range Ambush Platoon: Heavily-armed, Ranger/173rd AB volunteer, duty-specific, provisional unit; aka SRAP

Short Round: Artillery, mortar, rocket or naval shell that impacts before it reaches all the way to its intended target

Short Time / Long Time: Time length for which a prostitute is hired and figures heavily in payment negotiation

Short: A happy term denoting that a person has little time left in his tour of duty, usually less than 100 days

Short-Timer: Denotes the person nearing the end of a tour of duty; see also Short or Double- or Single-Digit Midget

Short-Timer's Calendar: Numbered-section drawing which is shaded each day to keep track of time left in the war

Short-Timer's Stick: 30-notch twig that one-a-day was broken off until one is left with only the stub – on his last day

Shot!: Radio code that artillery rounds are on their way and that friendlies need to seek cover immediately

Shotgun: An armed individual who guards a vehicle from either inside or riding on it; the right front vehicle seat

Shove Off: Lingo for to leave; go away

Shrapnel: Fragmented pieces of a bomb, shell, rocket, or other munition that are violently expelled as it detonates

Shrike: AGM-45A SAM-site killer, air-to-surface, radar homing, 390lb, 8" x 10ft, missile with a 15mi range

SI: Medical classification meaning *Seriously Ill*; see also VSI

Sick Bay: Navy clinic; see also Dispensary

Sick Call: A visit to Sick Bay for some ailment

SID: Seismic Intrusion Device; usually a dart-like, air-dropped, camouflaged, transmitter; see also TurdSID

Sidewinder: Raytheon AIM-9 short-range, air-to-air, infrared homing, 188lb, 5" x 9'11", Mach 2.5 missile

Sidgee: Slang for CIDG or one of its operatives

Sihanouk Trail: Glib term for the Ho Chi Minh Trail within Cambodia due to the "neutral" King's implicit approval

Simmons Rig: Linked metal ladder hung from a helicopter that troops could grab and be lifted out of the jungle

Sin Loi: See Xin Lỗi

Single-Digit Midget: Lingo for a person with less than 10 days left in his tour of duty; Double-Digit Midget and Short

SitRep: Situation Report; a radioed short narrative from field units or perimeter bunkers to command operators

Six: Lingo for at my rear or behind me; directionally 6 o'clock on a watch

Six: Radio designator for any company commander or that of a higher unit

Six-By: A ten-wheeled, three-axle, flatbed, 2½-ton truck with several variants; see also Deuce-And-A-Half

Sixty: An M60 machinegun or an M2 60mm mortar

Skank: Slang for a dirty, trashy female

Skate: Lingo for an individual who is a goof off; to slide by without doing any work

Skimmer: Lingo for a riverine 13.3ft Boston Whaler; outboard-motor, fiberglass boat

Skinny: Intelligence or information; see also Dope and Intel and Poop and Scoop and Word

Skipper: A ship's captain or commanding officer; affectionate nickname for a Marine Captain

Skirmish: A usually unintended, short, violent encounter between small elements of belligerent forces

Skivvie Honcho: Slang for a playboy or Butterfly

Skivvie House: Slang for a brothel or whorehouse

Skivvies: Lingo for underwear

SKS: Soviet 7.62mm, semi-automatic, 10-rd, 1,000-meter range, infantry rifle, precursor to the AK-47

Sky Crane: See CH-54

Sky Out: Slang for get out quickly

Sky Pilot: A Chaplain; term made popular by the rock music group *The Animals* from their song by the same name

Sky Soldier: A member of the 173rd Airborne

Sky: Slang for leave

Skyhawk: See A-4

Skylarking: Lingo for not doing one's job; goofing off; not putting in the effort

Skymaster: See O-2

Skyraider: See A-1

Skyspot: Ground-Directed Bombing; aka Combat Skyspot; see GDB

Skytrain: See C-47

Slack Man: The second man in a column of infantrymen on a combat patrol; aka Deuce Point

SLAM: Seeking Locating Annihilating & Monitoring; coordinated, massive all-asset, bombing attack, including B-52's

Slant: Derogatory term referring to the enemy and sometimes all Vietnamese; see also Dink, Gook, Gooner, Slope

Sleeper: Lingo for an individual who is a covert agent or mole

SLF: Special Landing Force; amphibious light infantry air-ground team embracing flexibility, surprise, and maneuver

Slick Sleeve: Navy individual w/no hash mark on his uniform, meaning he has less than 4-years of unbroken service

Slick, Medevac: See Dust Off

Slick: UH-1 lightly-armed (usually only two M60 machineguns) troop/cargo helicopter with no heavy guns or rockets

Slick: AF Huey fitted with two M60 machineguns and a rope ladder used for emergency extraction of SF troops

Sling Load: A load hung under a helicopter via straps or net or cable

Slop Chute: Lingo for a bar

Slop Chute: Navy term for a sluice at the aft of a ship used to dump garbage

Slope: Derogatory term referring to the enemy and sometimes all Vietnamese; see also Dink, Gook, Gooner, Slant

Slow Mover: Lingo for a propeller driven fixed-wing aircraft; see also Fast Mover

Smack: Slang word for heroin

Small Arms: A weapon, not crew-served, that is portable and can be fired by an individual without assistance

Smart Bomb: Laser-guided, gravity munition fitted with controllable fins from a Shrike missile and a guidance kit

Smell Good: A slang term meaning cologne or deodorant

Smell-O-Meter: Slang for a People Sniffer

Smith-Corona Commando: See Remington Raider

Smokey The Bear: Sardonic term for Army Drill Sergeants who wear campaign hats

Smoking Lamp: Decree for when smoking is authorized; if lamp is lit, one can smoke; if out, one cannot smoke

SNAFU: Situation Normal, All Fucked (Fouled) Up

Snake 'N Nape: Slang for 500lb Snake Eye bombs and 500lb napalm cannisters; see also Shake 'N Bake

Snake Eye: Lingo for 500lb bomb fitted w/pop-out fins, delivered as close air support by low-level fighter-bombers

Snake Eye: See MARK-15 Retarding Device

Snake: Slang for an AH-1 Cobra helicopter gunship; aka Cobra or Huey Cobra

Snap In: The practice of dry-firing a weapon, usually a handgun or rifle

SNCO: Staff Non-commissioned Officer; enlisted rank E-6 through E-9; see addendums

Sneaky Pete: Affectionate term referring to a Ranger or a Special Forces soldier

SNIE: Special National Intelligence Estimate; CIA-prepared secret reports meant for only the very highest authorities

Sniper: Highly-accurate and stealthy rifle expert who shoots from cover at the enemy, usually from long ranges

Snipes: A derogatory Navy term for a ship's engine room sailors; aka Bilge Rats

Snooping & Pooping: A reconnaissance patrol searching (*Snooping*) and collecting information (*Pooping*)

Snoopy: A helicopter flies low-level attempting to draw fire while a higher aircraft attacks the source; see Fire Fly

Snoopy: Helicopter fitted with vacuum sniffers to detect urine odors to discover massed enemy in the jungle

Snot Locker: Slang for your nose

Snuffy: Slang for lower-ranked enlisted person; sometimes used to describe an infantryman

Soften An LZ: The bombardment of a suspected perilous landing zone prior to helicopter touchdown; see also Prep

SOG: Studies and Observations Group; covert teams specializing in deep penetration of enemy territory

SOI: Signal Operating Instructions; a 1st Signal Brigade printed guide with classified frequencies and call-signs

Soldier: A member of the U.S. Army

Solid: Slang for a favor; excellent

Song: Vietnamese word for river

SOP: Standard Operating Procedure

Sorry `Bout That!: Sarcastic term that one uses when not actually remorseful

Sortie: Aviation term for one aircraft conducting one mission, multiple missions = multiple sorties

SOS: See Shit On A Shingle

Soul Brother: Slang for an African-American

Sound Off: Lingo meaning *speak up!* or shout out a specific phrase in formation or use a command voice

Soup Sandwich: Slang for a sloppy outcome or disheveled appearance or disorganized or worthless or hopeless

Southeast Asia Treaty Organization: International org for the defense of SEA, but without military clout: see SEATO

Southeast Asia: The geographical area consisting of Vietnam, Laos, Cambodia, and the surrounding waters

Soviet: General perception of a citizen or anything made in the Union of Soviet Socialist Republics; aka USSR

SP Pak: Sundry Package sent to field troops; pens/paper, candy, soap, razor blades, toothbrushes, +; aka Benny Pak

SP: Shore Patrol; Navy designation for its military police

SP: Start Point; the beginning of something

Space Cadet: Slang for a person who is routinely unmindful or unaware of what is happening around him

Spaced Out: Slang term meaning a state of non-functionality due to excessive drug use; see also Trippin'

Spaced Out: State of mind when one is daydreaming or in deep thought

Spad: See A-1

Spade: Denigrating slang term for an African American

Sparrow Hawk: Lightly-armed, USMC infantry platoon with two, three, or four helicopters standing by as an RRF

Spaz: Slang for a clumsy or dimwitted person

Spear Chucker: Derogatory term used by Whites toward Blacks

Special Forces: See Green Beret

Special Purpose Insertion & Extraction Rig: See SPIE Rig

Spectre: Codename for AC-130 gunship with two 40mm autocannons, a 20mm cannon, and a 105 howitzer

Spider Hole: Lingo for a small, cleverly camouflaged, enemy foxhole

SPIE Rig: Long, heavy strap with woven loops for Recon Marines to D-Ring clip onto & be heli-lifted from the jungle

Spikebouy: A dart-like, air-dropped, camouflaged, movement or sound sensing, transmitter; see SID

SPIW: Special Purpose Individual Weapon; Army program for a workable flechette-firing rifle and other concepts

Splash: Radio code that an artillery marker round has impacted, usually followed by adjust-fire requests

Splib: Non-derogatory racial reference used by White Marines to denote a Black Marine

Split The Scene: Slang for depart from the location

Split: Slang for leave; depart

Sponge: Slang for a freeloader; a taker, not a giver

Spook: Lingo for a spy or covert operative or cryptographer or a person in the intelligence services

Spooky: See PUFF

Spotter Round: Single round revealing its impact with smoke or fire to adjust for barrage; aka Marker/Target Round

Spud Locker: Navy lingo for a pantry in a Galley

Squad Bay: Navy term for a wide open Barracks area

Square: Slang for a person who is uncool or does not fit in with group norms

Squared Away: Lingo for an organized person, shop, unit, etc. that is neat and orderly

Squelch Grease: A fruitless-search prank on FNG's sent to find this non-existent item

Squelch: A electronic suppressive measure to reduce the hiss noise on a communications radio

Squid: Insolent term for a Sailor

Squids Are For Kids: A mocking phrase patterned after the popular TV ad *"Trix Are For Kids"*

SR-71: Lockheed *Blackbird*; Mach 3.2, 85,000ft ceiling, tandem seat, long-range, strategic reconnaissance aircraft

SRAO: Supplemental Recreational Activities Overseas; aka American Red Cross

SRAP: See Short Range Ambush Platoon

SRO: Senior Ranking Officer; the military person with the highest echelon in a POW camp

SSS: Selective Service System; see addendums

STAB: See SEAL Team Assault Boat

STABO: Stabilized Body rig; extraction/rescue harness that one can strap onto and be air-lifted out of the jungle

Stack Arms: The act of placing three rifles butt-down and muzzle-up, in a vertical, conical, self-supporting pile

Stacking Swivel: A Drill Instructor's sarcastic term meaning your neck

Stacking Swivel: Projection under a rifle muzzle used to Stack Arms

Stand Fast: An order to remain in place without moving around; to stay still

Standby: Wait; pause until further instruction

Stand-Down: An official period of inactivity from fighting, usually for resting, refitting, or training

Starboard: Navy term for the right when facing forward

Starchies: Lingo for starched Utilities

Starlifter: See C-141

Starlight Scope: AN/PVS-2 Green Eye night-vision device; uses moon and stars to intensify/enhance observation

Starlite: First major ground combat operation (18-24 AUG 1965) in SVN, fought by USMC on the Batangan Peninsula

Stay-Behind: Deceptive tactic of leaving a squad hidden in place to ambush the enemy after the larger unit departs

Steam & Cream: Steam rooms with massages and happy endings (ejaculations) provided by Vietnamese females

Steamer: An establishment for Steam & Creams

Steel Pot: See Helmet

Stern: Navy term for the back of a boat or ship

Stinger: Fairchild AC-119; see also Flying Boxcar

Stinger: Lingo for the tail skid at end of a helicopter tail boom that prevents the tail rotor from hitting the ground

Stingray: A tactic using Recon assets offensively on targets of opportunity, rather than for intelligence-gathering

Stockade: Army term for a jail; confinement area

STOL: Short Takeoff and Landing

Stoner: Slang for a person who uses illicit drugs, especially marijuana

STRAC: Complimentary lingo meaning *Strategic, Trained, and Ready Around the Clock*

STRAC: Lingo for Strategic Air Command

STRAC: Satirical term for *Shit, The Russians Are Coming*

Strack: Army lingo meaning the best; outstanding

Straight Arrow: A person who remains faithful to his wife (girlfriend), his service, or his faith

Straight Leg: An infantryman who is not airborne-qualified; aka Leg

Straphang: To be OPCON'ed to another team or unit

Straphanger: Airborne term for a person who is not assigned to a unit, but is along for the ride or attack

STRATA: Short-Term Reconnaissance And Target Acquisition; quick cross-border intelligence gathering

Strategic Air Command: See SAC; aka STRAC

Strategic Hamlet Program: Village self-defense plan that reconstructed hamlets into buttressed compounds

Stratofortress: See B-52

Stratotanker: Boeing KC-135; 4 turbojet, fixed-wing, Air Force inflight refueling aircraft; aka Gas Station In the Sky

Street Without Joy: Route 555; parallel between Highway 1 and the South China Sea southeast of Quang Tri City

Street Without Joy: The 1961 best-selling book by Bernard Fall detailing the horrors of the French Indochina War

Stretcher: A Litter on which to carry the wounded or sick; see also Poncho Litter

Strike Force: Platoon-sized, highly-mobile, heavily-armed, Search & Destroy team

Strobe: Flashing light to mark locations, to signal aircraft, or to act as a warning

Stroke Book: Slang for a pornographic magazine

Strongpoint: Difficult to overrun/avoid, easily-protected, usually on high ground, hardened defensive position

Submachine Gun: A short-range, close-quarters fighting, automatic rifle; see Thompson and Grease Gun

Sucking Chest Wound: Thorax wound through which air enters the lungs rather than through the trachea

Sugar Report: Slang for a letter from a girlfriend, child, or wife

Sundry Pack: See SP Pak

Super Jolly Green Giant: Sikorsky HH-53; purpose-designed, single-rotor, twin jet-engine, combat SAR helicopter

Super Sabre: See F-100

Supply Classes: See addendums

Suppressive Fire: Counter-munition shooting in such violence and intensity as to overwhelm and stop enemy guns

Survey: Navy term meaning to get rid of or to turn in an unserviceable item

SVN: South Vietnam

Swab: Mop

Swabbie: Insolent term denoting a Sailor

Swift Boat: Many-missioned, 50ft, 25kph, aluminum, shallow-draft, combat watercraft; aka PCF (Patrol Craft, Fast)

Swiss Seat: Extremely uncomfortable rappel rope harness for use in emergencies; see McGuire Rig

Syrette: A small, disposable, single pre-dosed (usually morphine) hypodermic needle, squeezable-tube, injector

T

T&T: Wound code meaning a Through & Through wound; a projectile passing entirely through a person

T-54 / T-55: Soviet, 35-ton, 31mph, 317mi range, 100mm gun, medium battle tank series

TAC(A): Tactical Air Controller (Airborne); the pilot or AO responsible for airstrike and artillery target management

TAC: Tactical Air Command; responsible for all Air Force operations in SVN, NVN, Laos, and Cambodia

TACAIR: Any aircraft in active support of ground troops in the fight

Tactical Area Of Responsibility: See TAOR

Tactical Zone: See addendums

TAD: Temporary Additional Duty; aka TDY

TAD: *Travel Around Drunk*

Tail Boom: Narrow tube-like structure extending from a helicopter fuselage that holds the tail rotor assembly

Tail: See Tail-End Charlie

Tail-End Charlie: The last man in a column of infantrymen on a combat mission

Take a Dump: Slang for to defecate

Take a Leak: Slang for to urinate

Tally Ho: Radio code from a pilot that he has his target in sight and is starting his attack

Tanglefoot: Single-strand barbed wire laid in a mesh pattern at ankle height, used to ensnare infiltrating enemy

Tango Boat: Armored troop carrier, shallow-draft, twin-diesel, modified 57ft LCM, riverine assault boat; aka ATC

Tango Uniform: Military phonic alphabet for Tits Up meaning ruined or broken or dead

TAOR: Tactical Area Of Responsibility; geographically-designated region assigned for a military unit to control

Tapped Out: Slang for having no money left or having nothing left of value

Taps: Lights out; end of day ritual bugle signal; also played at military funerals and memorial services

TARFU: *Things Are Really Fucked (Fouled) Up*

Target Round: Single round revealing its impact with smoke or fire to adjust for barrage; aka Marker/Spotter Round

Tarhe: See CH-54

Tarmac: Any area of an airfield that is paved whether with asphalt, concrete, or other hard-packed material

Task Force 77: The 7th Fleet's carrier-based, strike operations from the South China Sea or Tonkin Gulf

TDY: Temporary Duty; aka TAD

Tear Gas: See CS

TEC: See Tail End Charlie

Tee-Tee: Bastardized Vietnamese word for small or little or tiny; aka Ti-Ti

Ten-Percenter: Individual who is one of the few Non-Hackers in a unit

Tet Offensive: 1968 NVA & VC coordinated attack on 100 SVN towns/cities in unfulfilled hopes of a general uprising

Tet: A 5-day holiday celebrating the Vietnamese Lunar New Year

Teufelhunde: Reference to Marines and the ferociousness of mountain dogs in Bavarian folklore; see Devil Dog

TF: Task Force; a unit temporarily, or provisionally, formed to conduct a particular mission

TFES: Territorial Forces Evaluation System; same as HES, but for analysis of militia forces level of readiness

TFR: Terrain Following Radar; aircraft all-weather radar that maps terrain ahead and warns of impending hazards

The Last Supper: Lingo for the final meal before taking the Freedom Bird

There It Is: Slang phrase meaning *yes* or *I agree* or *exactly* or *I told you so*

Thermite: Non-explosive, easily-ignited, anti-materiel incendiary that burns at extremely high temperature

This Is No Shit: Slang preparatory phrase used when something exaggerated or untrue is about to be said; aka TINS

Thompson: An M1921 11" barreled, .45cal, 725 rounds-per-minute Submachine Gun; aka Tommy Gun

Thousand-Yard Stare: A battle-worn, traumatized soldier's unfocused gaze

Threads: Slang term for clothes

Three Hots & A Cot: What grunts looked forward to after returning from the bush; 3 hot meals & a place to sleep

Through & Through: See T&T

Thud: See F-105

Thumper: See M-79

Thunder Road: Nickname reserved for very dangerous thoroughfares

Thunderchief: See F-105

TI: Training Instructor in the Air Force; aka Military Training Instructor

Tiger Balm: Eucalyptus-based, homemade, analgesic salve; supposedly eases pain and cures a multitude of ailments

Tiger Piss: Any poor-tasting beer; also Vietnamese *LaRue* brand beer with a tiger on its label

Tigers: Unofficial Tigerstripe, camouflage fabric used by special ops due to its close-range, concealment properties

Tigerstripe: A tight-pattern, camouflage fabric usually issued to SVN troops; also worn by U.S. SF; see also Tigers

Tight: A slang reference to the strong bond of close friends in agreement and purpose; well-functioning

Time On Target: See TOT; Note: This is for artillery

Time Over Target: See TOT; Note: This is for aircraft

Time-In-Grade: Policy to slow rapid promotion by requiring a specific total of months in a rank; see Shake 'N Bake

TINS: See This Is No Shit

Tipsy 25: AN/TPS-25 doppler ground-surveillance radar detecting movement direction/numbers/coords of enemy

Ti-Ti: See Tee-Tee

Tits Up: Ruined; broken; dead; aka Tango Uniform

To The Colors: Bugle call for the raising or striking (lowering) the flag; see also Colors

TO: Table of Organization; hierarchal listing of staffing, duties, authorized equipment, and status of a unit

TOC: Tactical Operations Center; liaison between infantry, artillery, air, and other assets in a battle situation

TOE / TO&E: Table of Organization & Equipment; see TO

Toe Popper: Lingo for a Special Forces tiny disk-shaped (2.2"dia x 1.5"), 3.5oz, anti-personnel mine; aka M14

Tommy-Gun: See Thompson or M1921

Tonkin Gulf Incident: 2 AUG 1964 attack that initiated U.S. entry into the war; see also C. Turner Joy and Maddox

Tonkin Gulf Yacht Club: Facetious term for the U.S. Navy in Vietnam

TOO: Targets Of Opportunity

Top: Lingo for *First Sergeant*; see also First Shirt and addendums

Topside: Navy term for upstairs, or on an open deck

TOT: Time On Target; when rounds of various trajectories are fired in sequence to all hit a target simultaneously

TOT: Time Over Target; the moment when an aircraft is scheduled to attack (or photograph) a target

Touch & Go: The training exercise in which an aircraft lands and then immediately takes off again; aka Crash & Dash

Tour Of Duty: Length of time assigned In-Country; generally 12mo (13mo for Marines), subsequent tours were 6mo

TOW: Tube-launched Optically-tracked Wire-guided anti-tank missile system; aka BGM-71

Tracer: A visible color streak left momentarily in the air to track the path of a bullet (green=communist; red=NATO)

Track: Lingo for any vehicle that moves on linked tracks, not wheels

Tractor Rat: Nickname for a member of an AmTrac unit

Tractor: An in-unit nickname for an AmTrac; see LVT

Trajectory: The arc of a projectile through the air

Transitional Lift: Act of a helicopter as it adapts from hover to forward flight so the rotors can grab undisturbed air

Transport Pack: Marching Pack plus the Haversack

Traumatic Amputation: Sudden, violent loss of an extremity, including the head, by explosion, gunfire, or shrapnel

Tree Burst: The inadvertent detonation of a shell hitting trees or tall objects before reaching the intended target

Tree Line: Tall, usually dense vegetation that grows at the edges of fields, paddies, and open areas; aka Wood Line

Trench Foot: See Immersion Foot and Paddy Foot and Jungle Rot

Trench Monkey: Disparaging lingo meaning a U.S. Army infantryman

Triage: Sorting and allocating treatment to the wounded based on the seriousness of their wounds vs survival odds

Tri-Border: The confluence where Vietnam, Cambodia, and Laos intersect

Trigger Puller: Friendly term for an infantryman

Trioxane: The flammable material making up a Heat Tab

Trip Flare: M49A1 35,000 candlepower, magnesium, instant-firing, tripwire-triggered, perimeter defense flare

Trip Wire: Monofilament/string/vine/wire attached to a boobytrap and strung across where a man is likely to walk

Triple A: Anti-Aircraft Artillery; aka AAA

Triple Canopy: Jungle growth of multiple, vegetative layers at ground/intermediate/high levels that block most light

Triple S: Sardonic term for the morning routine of Shit, Shower, and Shave

Trippin': Slang for inebriated, unawareness state of a person on drugs; aka Spaced Out or Zoned Out

Tropic Lightning: Nickname for the U.S. Army's 25th Infantry Division

Tropical Hershey Bar: Essentially a non-melting chocolate treat in some sundry packs; see Montagnard Bar

Tropo: Lingo for huge antennas that provided long-range communications by bouncing signals off the troposphere

Trops: Summer (Tropical) Class-A, khaki, dress uniform

Truck War: Lingo for the bombing campaign against vehicle traffic on the Ho Chi Minh Trail

Truong Son Corridor: Enemy supply line in SVN that paralleled that of the Ho Chi Minh Trail in Laos, Trường Sơn

TTY: Teletype

Tu Do Street: Saigon street infamous for its prostitution, bars, crime, and shops

Tube of Super Torque: A fruitless-search prank on FNG's sent to find this non-existent item

Tube: Lingo for the barrel of an artillery piece, mortar, or rocket launcher

Tunnel Rats: Soldiers, usually of small stature, sent down into tunnels to search them and kill or capture the enemy

TurdSID: Lingo for a feces-looking, dropped, Seismic Intrusion Device that detects and reports vibration/sound data

Turn To: Lingo meaning *Get to work*

Turns: Lingo for the rpm of helicopter rotors

Turret Tops: Condescending term for personnel assigned to tank crews

Turtles: Lingo for replacement troops, so-called for the assumption that they took their sweet time getting there

Two Shop: Lingo for G-2 or S-2 Intelligence elements

Two-Stepper: See Banded Krait

Type 59: Chinese 35-ton, 31mph, 273mi range, 100mm main gun, main battle tank

Type 62: Chinese 21-ton, 37mph, 310mi range, 85mm main gun, light tank

Type 63: Chinese 18-ton, 40mph (7mph hydrojet), 230mi range, 85mm main gun, light amphibious tank

U

U-1A: See DHC-3

UA: Unauthorized Absence; any absence from a post or duty with no intent to return; see also AWOL

UCMJ: Uniform Code of Military Justice; the foundation of military law

UFN: Until Further Notice

UH: Utility Helicopter

UH-1: Bell *Iroquois* series of variants of its utility helicopter; see also Huey and Huey Slick

UHF: Ultra High Frequency

UN: United Nations

Un-Ass: Lingo for straighten out oneself, or to fix something or to get a better attitude

Un-Bloused: Trousers that are not tucked into boots

Uncle Ho: Depending on one's perspective, an affectionate, or mocking, term for Ho Chi Minh

Uncle Sam: A mocking term for an Army draftee referring to the prefix *US* on their service number

Under Arms: Carrying or wearing a weapon

Under Fire: The condition in which the enemy is shooting at or bombarding you

Unfuck Yourself: Lingo for straighten out oneself, or get a better attitude or an order to do something better

Unglued: Slang for falling apart emotionally, or distraught or ranting or angry

Union of Soviet Socialist Republics; aka USSR

Unit One: Corpsman medical kit; included surgical kit, combat dressings, morphine, Darvon, etc.; see also M-3 / M-5

Universal Coordinated Time: See Zulu Time

Unreal: Slang for absurdly ridiculous, or unbelievable, or praise for some great thing

Upper Echelon: Snuffy's reference to those in control

Uptight: Slang for tense or on edge, or nervous, or not cool

US Army: A soldier's sardonic term meaning *Uncle Sam Ain't Released Me Yet*

US Army: A soldier's sardonic term meaning *U Sons-a-bitches Are Ruining My Youth*

US: United States

US: A sarcastic Army draftee term meaning *Unwilling Service*, because of the prefix *US* on their service number

US: Army draftee designation noted by the prefix US on one's service number; see also RA

USA: United States Army

USA: United States of America

USAF: Air Force sardonic term meaning *U Sure Are Fucked*

USAF: United States Air Force

USAID: U.S. Agency for International Development; administers foreign aid and enhances socio-economic progress

USARV: United States Army, Republic of Vietnam; the highest-level Command of the U.S. Army in SVN

USCG: *United States Coast Guard*

USELESS: Derogatory term for the public relations office USIS

USIS: United States Information Service; PR Office that highlighted U.S. views and weakened those of the USSR

USMC: Marine sardonic term meaning *Uncle Sam's Misguided Children*

USMC: Marine sardonic term meaning *Uncle Sam's Mountain Climbers*

USMC: Marine sardonic term meaning *Useless Suffering and Meaningless Chaos*

USMC: United States Marine Corps

USN: United States Navy

USO: United Service Organization; provided morale-raising entertainment, goodies, and other items for the troops

USSR: Union of Soviet Socialist Republics; see Soviet

Utilities: The standard work/combat olive drab uniform for the U.S. Marine Corps

UVG-0029: See Rubber Bitch / Rubber Lady

V

V: Short for V-100

V-100: Cadillac Gage M706 *Commando* armored vehicle; 11-ton, 62mph, used for convoy escort/airbase security

V40: See Mini Frag

VA: Veterans Administration

VC: Military phonic for Victor Charlie, aka Viet Cong or Cong for short; see also Chuck, Charlie, and Mr. Charles

VCF: Viet Cong, Female

VCI: Viet Cong Infrastructure; shadow government poised to take over all offices and power when the war ended

VCS: Viet Cong Suspect

VE: Inserting helicopter troops/surrounding the enemy/forcing it to redirect its forces to fight a 360-degree battle

Vertical Envelopment: See VE

VFR: Visual Flight Rules; using eyes to maintain headings/avoid obstacles in good visibility conditions; see also IFR

Victor Charlie: See VC

Viet Cong: Vietnamese Communists; indigenous South Vietnamese who believed in and fought for totalitarianism

Vietnam Era Veteran: Those who served in the armed forces during the Vietnam years, but never In-Country

Vietnam Veteran: The accepted term for anyone who was stationed In-Country during the war

Vietnamization: President Nixon's *Keystone* program to slowly phase out U.S. troops and turn over the war to SVN

Vil or Ville: A Vietnamese village or hamlet

VIP: Very Important Person

VN: Vietnam (South)

VNAF: Vietnam (South) Air Force

VNMC: Vietnam (South) Marine Corps

VNN: Vietnam (South) Navy

Volley: See Salvo

Voodoo: See RF-101C

VR: Visual Reconnaissance; low-level, low speed flight mission over a previously uncharted and likely enemy area

VSI: Very Seriously Ill; sickness degree code for when imminent death is likely without immediate care; see also SI

VTR: Vehicle, Tracked, Recovery; see M88

Vulcan: General Electric, 6-barrel, air-cooled, burst-limited 6,000 round-per-minute, 20mm rotary cannon; aka M61

W

WAAPM: Wide Area Anti-Personnel Munition; see Cluster Bomb

WAG: Lingo for *Wild-Ass Guess*

Wait-A-Minute Vine: A foot-entangling creeper; so-called because that was what one would say when snagged

Wake Up: Lingo for one's last morning in Vietnam

Walking Barrage: Artillery fire first laid close to friendlies and then slowly moving the impacts nearer to the enemy

Walking Wounded: Combat injured persons who are able to walk despite their wounds

War Of Attrition: To destroy enemy troops/supplies faster than can be replaced so the enemy cannot wage war

War Powers Act: November 1973 law limiting a U.S. president from sending troops abroad to no more than 90 days

Warrant Officer: Highly-skilled technician, usually a helicopter pilot, ranked above SNCO, but below officer status

Waste: Lingo for to kill or destroy; see also Blow Away or Grease or Hose or Nail or Zap

Wasted: Slang term meaning a state of non-functionality due to excessive drugs or alcohol

Watcher: See LZ Watcher

Water Boo: Slang for Water Buffalo

Water Buffalo: Horned, domesticated, beast of burden that can grow to 2,200 pounds, used to plow and pull carts

Water Buffalo: Lingo for a large portable water tank

Wave Off: Aborted landing

Weapons Weenie: A derogatory term for a weapons mechanic/loader; aka Load Toad

Web Gear: See M-1956

Weed: Slang for marijuana; see also Grass or Dope

WESTPAC: Term for Western Pacific (not the railroad company); usually meant service in Vietnam

Wet Rations: See C-Rations

What's Your Bag: Slang for *What do you do for a living?* or *What's wrong with you?*

Whiskey Papa: Military phonic alphabet for White Phosphorus

Whispering Death: NVA's nickname for B-52 carpet bombing that could not be seen nor heard until it was too late

White Mice: Mocking term for the SVN traffic police, due to their white shirts, helmets, and gloves

White Phosphorus: Volatile clay-like chemical that burns at 5,000°F when exposed to air, used in many munitions

White Walls: Lingo for the standard Marine Corps haircut; see High & Tight

WIA-E: Casualty classification for *Wounded in Action, Evacuated*

WIA-NE: Casualty classification for *Wounded In Action, Not Evacuated*

Widow-Maker: Slang for the M-16 rifle

Wild Weasel: Code name for any aircraft fitted with anti-radar missiles and a mission of killing radar and SAM sites

Willie Pete / Willie Peter: Affable nicknames for White Phosphorous

Wilson Pickett: Lingo for White Phosphorous

Windage: A rifle-sight tweaking to correct the adverse effect of wind forces on a projectile's intended path

Wipe Out: Slang for an accident

Wire: Barbed or concertina variants strung around a perimeter and used as an obstacle to deter an attacking enemy

Wizz: To urinate

WO: Warrant Officer

Wobbly One: Lingo for a WO-1, a brand new Warrant Officer

Wood Line: See Hedge Row or Tree Line

Word, The: Intelligence or information; same as Dope and Intel and Poop and Scoop and Skinny

World Of Hurt: Lingo for what the Drill Instructor puts you in; corporal punishment for a serious transgression

World, The: Slang for the United States

WP: White Phosphorus

WX: Weather

X

Xin Lỗi: Apologetic Vietnamese term for *Sorry,* pronounced Sin Loi

XM177: *Colt Commando;* see CAR-15

XO: Executive Officer; the second in command of a company or higher unit, usually more admin than command

Y

Yankee Station: Code name for the U.S. Naval operational area in the Tonkin Gulf; see also Dixie Station

Yard: Lingo for Montagnard (in French: mountain person); member of one of 54 aboriginal tribes found in Vietnam

YAT-YAS: USMC AmTrac lingo for *You Ain't Tracs, You Ain't Shit*

Yellow Brick Road: Sardonic term for the Ho Chi Minh Trail

You'll Wonder Where the Yellow Went When Napalm Hits the Orient: Cynical play on a popular toothpaste jingle

Young Tigers: Nickname for KC-135 *Stratotanker* crewmen who refueled fighter-bombers and B-52's mid-air

Your Shit Don't Stink: The state of one's belief that they are personally better than, more astute than, anyone else

Z

Zap: Lingo for kill or destroy; see also Blow Away or Grease or Hose or Nail or Waste

Zero-Dark-Thirty: Lingo meaning pre-dawn hours; very early

Zero-Zero: Aviation term meaning Zero ceiling (ground-level clouds) and Zero forward visibility; see IFR

Ziggy: Lingo for to move quickly in a different direction or directions to evade what's attacking you; see also Jink

Zippo Boats: See Zippo

Zippo Raid: The burning of enemy villages or of VC sympathizer villages

Zippo Squad: A flamethrower team

Zippo Track: See Zippo

Zippo: A brand of popular cigarette lighter often engraved with some war-themed epitaph

Zippo: Nickname for a vehicle/craft fitted with a flamethrower; see also M132 and M67 and PGM-5 and M2A1-7

ZOA: See Zone Of Action

Zone & Sweep: A 30-round artillery barrage at target center, plus one on all 4 sides to form a "+" or "X" pattern

Zone Of Action: A tactical subdivision or an TAOR that applies to an attack or retrograde; aka ZOA

Zoned Out: Slang for a state of mind when one is daydreaming or staring in deep thought; see also Trippin'

Zoomie: Mildly-affectionate term for a jet aircraft pilot; aka Jet Jockey

Zoot Suit: Sardonic term for a flight suit

Zulu Time: 0° longitude or Universal Coordinated Time; zero begins with Z, see phonic alphabet in addendums

Zuni: Folding-Fin Aerial Rocket (FFAR); 5-inch x 77-inch long, 79.5lb, pod-fired, unguided, air-to-ground missile

0 – 8

0311: The Marine Corps infantry rifleman MOS; see also O-3

105: Lingo for a 105mm Howitzer

105: See F-105

105mm: Rock Island Arsenal M102; 105mm towed, howitzer; see also M108

106: See M40

10-Percenter: See Ten-Percenter

11 Boo: Lingo for 11-B

11-B: Army basic infantryman MOS

12.7: A .51 caliber communist heavy machinegun

120mm: Lingo for a Soviet mortar; see PM-43

122 mm: A 75.4" long, 101lb, fin-stabilized, tube or earth mound launched, Soviet rocket with a 6.5mi range

122: Soviet rocket; see 122mm

123: See C-123

130: See C-130

14: See M14

140: Soviet rocket; see 140mm

140mm: A 42.3" long, 90lb, spin-stabilized, tube or earth mound launched, Soviet rocket with a 6mi range

155: See M109

16: See M16

175: See M107

17th Parallel: Geneva Accords provisional border between NVN/SVN ending the French Indochina War; see DMZ

1-A: Conscription classification meaning the individual is fit for military service; see addendums

1-Holer: A single cavity over a pit into which one defecates; comes in several variants; see Shitter and Shithouse

1st Signal Brigade: The unified command for In-Country tactical and strategic communications

2.75: General Dynamics aircraft-mounted rocket diameter in inches; air-launched from pods

20, 20mm: See Vulcan

201 File: U.S. Army Personnel File; contains various documents pertaining to individual service

203: See M203 and SPIW

2-Holer: See 1-holer

2-Stepper: See Banded Krait

3.5: An M1 recoilless rocket launcher; aka bazooka; used early in the war; replaced by the M72 LAAW

33: Vietnamese beer; brewed using formaldehyde for clarity and to extend shelf life; aka Ba Moui Ba

3-Holer: See 1-holer

4.2: M2 4.2-inch heavy-weight mortar; aka Four-Deuce

40mm Bofors: Autocannon, 300-rounds per-min, 7½mi range, top-fed, anti-aircraft gun, used in Spectre; aka L/70

45 / .45: Generally, a .45 caliber automatic; See M1911A1

4-F: Conscription classification meaning the individual is unfit for military service; see addendums

4-Holer: See 1-holer

5.56: Size in millimeters of ammunition used in an M-16 infantry rifle; .223cal

50: See M2

51: A 12.7mm communist heavy machinegun

5-By, 5 X 5: Hearing Loud & Clear on a communications radio, if well

6 & A Kick: General Court-Martial punishment of 6-month pay loss, 6-months hard labor, and a DD; see addendums

6: Codename for an OH-6A helicopter

6: Designation for leader of any unit in a field operation

60: M19 (and later M225) 60mm lightweight mortar

60: See M60

7.62: This is a common 30-caliber NATO projectile (7.62x51mm) size used in the M14, M60, and the Minigun

782 Gear: Equipment such as ammo pouch, canteen, pack, etc. that is signed for on DOD Form 782; see Deuce Gear

79: Short for the M79 grenade launcher

81: M252, 81mm medium-weight mortar

82: Lingo for an 83mm Soviet medium-weight mortar; see PM-41

8-Inch: Eight-inch diameter artillery; see M110 and M115

8th & I: Nickname for *Marine Barracks* in Washington, DC; home to the CMC; aka Eighth & I

APPENDIX

WHY ADDENDUMS?

Rather than include equivalent-and-scattered entries in the glossary, they have been gathered into like-titled sections in the following pages for better context to one-another.

The military uses a phonic alphabet to make communications more easily understood, especially over a radio. Letter sounds are difficult to differentiate over a radio (bee, cee, dee, e, gee, pee, tee, vee, zee, etc.), particularly during stressful situations such as combat or emergency conditions or when there are background noises. Therefore, a system of phonics was developed. It resulted in a more reliable way of pronouncing letters.

The phonic alphabet has changed several times over the decades. Here is the one used in Vietnam:

ALPHABET – MILITARY PHONIC

(Radio Alphabet Code)

A – Alpha	N – November
B – Bravo	O – Oscar
C – Charlie	P – Papa
D – Delta	Q – Quebec
E – Echo	R – Romeo
F – Foxtrot	S – Sierra
G – Golf	T – Tango
H – Hotel	U – Uniform
I – India	V – Victor
J – Juliet	W – Whiskey
K – Kilo	X – X-ray
L – Lima	Y – Yankee
M – Mike	Z – Zulu

Chain of Command Ranks – Vietnam Era

		Air Force	Army	Marine Corps	Navy
OFFICER:					
	O-11	General of the Air Force	General of the Army	---	Fleet Admiral
	O-10	General	General	General	Admiral
	O-9	Lieutenant General	Lieutenant General	Lieutenant General	Vice Admiral
	O-8	Major General	Major General	Major General	Rear Admiral
	O-7	Brigadier General	Brigadier General	Brigadier General	Commodore
	O-6	Colonel	Colonel	Colonel	Captain
	O-5	Lieutenant Colonel	Lieutenant Colonel	Lieutenant Colonel	Commander
	O-4	Major	Major	Major	Lt Commander
	O-3	Captain	Captain	Captain	Lieutenant
	O-2	Lieutenant	Lieutenant	Lieutenant	Lt Junior Grade
	O-1	2nd Lieutenant	2nd Lieutenant	2nd Lieutenant	Ensign
WARRANT OFFICER:					
	W-4	---	Chief Warrant Officer-4	Chief Warrant Officer-4	Chief Warrant Officer-4
	W-3	---	Chief Warrant Officer-3	Chief Warrant Officer-3	Chief Warrant Officer-3
	W-2	---	Chief Warrant Officer-2	Chief Warrant Officer-2	Chief Warrant Officer-2
	W-1	---	Warrant Officer-1	Warrant Officer-1	Warrant Officer-1

ENLISTED:

Grade	Air Force	Army	Marine Corps	Navy
E-9	Chief Master Sergeant	Sergeant Major	Sergeant Major/Master Gunnery Sergeant	Master Chief Petty Officer
E-8	Senior Master Sergeant	First Sergeant /Master Sergeant	First Sergeant/Master Sergeant	Senior Chief Petty Officer
E-7	Master Sergeant	Sergeant First Class	Gunnery Sergeant	Chief Petty officer
E-6	Tech Sergeant	Staff Sergeant	Staff Sergeant	Petty Officer, First Class
E-5	Staff Sergeant	Sergeant/Specialist 5	Sergeant	Petty Officer, Second Class
E-4	Airman 1st Class/Sergeant*	Corporal/Specialist 4	Corporal	Petty Officer, Third Class
E-3	Airman 2nd Class/Airman 1st Class*	Private First Class	Lance Corporal	Seaman
E-2	Airman	Private E-2	Private First Class	Seaman Apprentice
E-1	Airman Basic	Private E-1	Private	Seaman Recruit

* Rank up to 1967 / Rank after 1967

85

In 1955, President Dwight D. Eisenhower established rules for behavior if, after all actions necessary to evade the enemy fail, an individual finds himself a prisoner of war (POW). These six principles explain that a prison camp is an extension of the battlefield. All servicemen were required to know these guidelines.

CODE of CONDUCT

ARTICLE I

I am an American fighting man. I serve in the forces which guard my country and our way of life. I am prepared to give my life in their defense.

ARTICLE II

I will never surrender of my own free will. If in the command, I will never surrender my men while they still have the means to resist.

ARTICLE III

If I am captured, I will continue to resist by all means available. I will make every effort to escape and aid others to escape. I will accept neither parole nor special favors from the enemy.

ARTICLE IV

If I become a prisoner of war, I will keep faith with my fellow prisoners. I will give no information nor take part in any action which might be harmful to my comrades. If I am senior, I will take command. If not, I will obey the lawful orders of those appointed over me and will back them up in every way.

ARTICLE V

When questioned, should I become a prisoner of war, I am bound to give only name, rank, service number, and date of birth. I will evade answering further questions to the utmost of my ability. I will make no oral or written statements disloyal to my country and its allies or harmful to their cause.

ARTICLE VI

I will never forget that I am an American fighting man, responsible for my actions, and dedicated to the principles which made my country free. I will trust in my God and in the United States of America.

In military terms, "Element" refers to an entity that is attached to a larger unit. Although there are variations in all elements except the individual by himself, the following is a general expectation of the named component, often wildly varying from branch-of-service to branch-of-service. Remember, this is the ideal, the striven for. In Vietnam, this prospect was invariably unrealistic.

ELEMENT SIZES

Unit	# of persons	Amalgamation of	Led by
INDIVIDUAL	1	1	Self
FIRE TEAM	3-4	3-4 Individuals	E-2 or E-3
SQUAD	8-14	3-4 Fire Teams + leader	E-4 or E-5
PLATOON	26-55	3-4 Squads + leaders	O-1 or O-2
COMPANY	80-250	3-4 Platoons + CP and attachments	O-3 or O-4
BATTALION	300-1,000	3-4 Companies	O-5
REGIMENT	1,000-3,000	3-4 Battalions	O-6
BRIGADE	3,000-5,000	2 Regiments	O-7 or O-8
DIVISION	6,000-20,000	3-4 Regiments or 2 Brigades	O-9 or O-10
CORPS	20,000-50,000	2-3 Divisions	O-10

An important part of military service is to share collective responsibility for security. Usually referred to as guard duty, there are two types of orders with which a guard must conduct themselves. One is Special Orders, which are distinctively situational. For all else, there are 11 General Orders that all military personnel need to know:

GENERAL ORDERS

1. To take charge of this post and all government property in view.

2. To walk my post in a military manner, keeping always on the alert, and observing everything that takes place within sight or hearing.

3. To report all violations of orders I am instructed to enforce.

4. To repeat all calls from posts more distant from the guard house than my own.

5. To quit my post only when properly relieved.

6. To receive, obey, and pass on to the sentry who relieves me all orders from the commanding officer, field officer of the day, officer of the day, and officers and noncommissioned officers of the guard only.

7. To talk to no one except in the line of duty.

8. To give the alarm in case of fire or disorder.

9. To call the corporal of the guard in any case not covered by instructions.

10. To salute all officers, and all colors and standards not cased.

11. To be especially watchful at night and during the time for challenging, to challenge all persons on or near my post, and to allow no one to pass without proper authority.

To maintain consistency in Command, Control, and Organization, the Basic Military Structure remains standardized throughout all command levels.

MILITARY STRUCTURAL LEVELS

DIVISION LEVEL and ABOVE

G-1: Personnel   (Basic staff system structure at the highest command echelons)
G-2: Intelligence   (")
G-3: Operations   (")
G-4: Logistics   (")
G-5: Plans   (")
G-6: Communications   (")
G-7: Training   (")
G-8: Finance; Contracts   (")
G-9: Civil Affairs   (")

BELOW DIVISION LEVEL

S-1: Personnel (Basic staff system structure at the subservient command echelons)
S-2: Intelligence   (")
S-3: Operations   (")
S-4: Logistics   (")
S-5: Plans   (")
S-6: Communications   (")
S-7: Training   (")
S-8: Finance; Contracts   (")
S-9: Civil Affairs   (")

SELECTIVE SERVICE SYSTEM

Conscription (The Draft) Classifications 1948-1976

1-A: Available for military service

1-AM: Medical specialist available for military service

1-A-O: Conscientious Objector; opposed to training/military service, available for noncombatant military service

1-A-OM: Medical specialist conscientious objector, available for noncombatant military service

1-C: Member of U.S. Armed Forces, the Coast and Geodetic Survey, or Public Health Service, or veteran

1-D: Member of a Reserve component or student taking military training

1-H: Registrant not currently subject to processing for induction or alternative service

1-O: Conscientious objector available for civilian work contributing to the national health, safety or interest

1-OM: Medical specialist conscientious objector available for civilian work contributing to the national health

1-S: Student deferred by status – (H) high school or (C) college

1-W: Conscientious objector performing civilian work in the national health, safety or interest – Released

1-Y: Registrant qualified for service only in time of war or national emergency; abolished in December 1971

2-A: Registrant deferred because of civilian occupation (except agriculture)

2-AM: Medical specialist deferred because of critical community need involving patient care

2-C: Registrant deferred because of agricultural occupation

2-D: Ministerial Students; deferred from military service

2-M: Registrant deferred for medical study

2-S: Registrant deferred because of activity in study

3-A: Hardship Deferment; deferred from military service because service would cause hardship upon his family

4-A: Registrant who has completed service; or sole surviving son

4-B: Official deferred by law

4-C: Alien or Dual National; sometimes exempt from military service

4-D: Ministers of Religion; exempted from military service

4-E: Conscientious objector opposed to both combatant and noncombatant training and service

4-F: Registrant not qualified for military service

4-FM: Medical specialist not qualified for military service

4-G: Sole surviving son/brother in a family where parent/sibling died in military or is in POW or M.I.A. status

4-W: Conscientious objector who has completed civilian alternate service

5-A: Registrant over the age of liability for military service

SOURCE: https://www.sss.gov/about/return-to-draft/

SOME THOUGHTS ON THE DRAFT

A notable change brought by President Nixon had been the fixing of inequities of the military Draft (which was renamed by FDR in 1940 as the congenial-sounding *Selective Service and Training Act* in anticipation of the U.S. entering WWII). During the Korean War, the system was modified when the government exempted college students who rated in the top half of their classes academically. Complaints that lower socio-economic, under-educated people were unfairly burdened by the Draft (while richer, college-attending individuals with deferments were not being conscripted in proportional numbers) were well-founded. Local SSS (Selective Service System) Boards were community-based and operated by local community members. Obviously, pressure from friends and powerful residents could skew and corrupt the system as to which boys would go and which boys would get to stay home. This system stayed in place throughout the Kennedy and Johnson presidencies.

To make the system more equitable, a lottery was instituted by Nixon in late 1969 which was based on birth date, rather than privilege. 366 capsules with birthdates (month and day) were mixed and drawn randomly. This made everyone of Draft age, who were ages 18-26 starting Jan 1, 1970, essentially equal. However, this system had unintended consequences. Since people of wealth, connections, and schooling were now suddenly eligible to be conscripted they became resentful, because here-to-fore they were quite content to have poor, non-college folks do their fighting and dying for them. There was a great backlash by the well-to-do, and protests against the war increased dramatically nationwide, especially on college campuses. The sheer hypocrisy of this change in attitude, from complacency to wrath, was lost on the privileged classes.

Graduates from the elitist universities did little to serve their country. Just 12 Harvard graduates died in Vietnam from 1962-1972. Princeton lost six and MIT lost two. By contrast, the nation's oldest military university, West Point, lost 333.

Throughout the war, 60% of eligible men avoided the Draft by qualifying for exemptions. 30,000 fled to Canada to dodge the Draft or went to Canada as deserters to evade Vietnam service. Ironically, an equal number of Canadians joined the U.S. military.

In the end, 73% of all those who died in Vietnam were volunteers. Two-thirds of all who served in the war were also volunteer citizen-soldiers.

STANDARDIZED CLASSES OF MILITARY SUPPLY

Class I: Rations; food and drinking water; free health and comfort items

Class II: Clothing & Equipment; individual equipment

Class III: Petroleum, Oil & Lubricants (POL) and other chemical products and coal

Class IV: Construction items; fortification and barrier materials

Class V: All types of ammunition, bombs, explosives, and supplementary items

Class VI: Health/hygiene items; and snacks, beverages, cigarettes, cameras, i.e. non-military sale products

Class VII: Large items like tanks, mobile machine shops, helicopters, vehicles, parachute systems, etc.

Class VIII: Medical equipment

> **Class VIII-a:** Medical consumables

> **Class VIII-b:** Blood & blood components

Class IX: Repair and maintenance parts and components

Class X: Non-military items to support agriculture and economic development programs

Class Miscellaneous: Captured material and salvaged items

TACTICAL ZONES IN SVN

South Vietnam was divided into four Military Regions. Collectively, they were Corps Tactical Zones and commonly referred to as follows:

I-Corps: This northernmost tactical zone consisted of five provinces. On the east, it bordered the Tonkin Gulf and the South China Sea. On the north, it bordered the Demilitarized Zone (DMZ) with North Vietnam. On the west, it bordered the country of Laos. The Roman numeral for 1 is always mispronounced "eye."

II-Corps: The next lower tactical zone consisted of 12 provinces and was broadly referred to as the Central Highlands. On its east was the South China Sea. Its northernmost western edge bordered Laos, but much of its western border was shared with Cambodia.

III-Corps: The next lower tactical zone consisted of ten provinces and surrounds the Capital Special Zone (Saigon and its outlying areas). Its eastern (southeastern) edge lies on the South China Sea. It's western edge borders Cambodia.

IV-Corps: This is the southernmost of the four tactical zones and known simply as The Mekong Delta, or just The Delta. On its east and south lies the South China Sea. Its northwestern edge borders Cambodia. On its western coast is the Gulf of Thailand.

TYPES OF DISCHARGES

Types of discharges have fluctuated to some extent over the years, but here is a synopsis for your perusal:

Honorable Discharge (HD): This is the highest of all discharges. It indicates professional conduct, faithfully executed duties, and otherwise admirable service. It is like being released from active duty with an A+.

General Discharge *Under Honorable Conditions* (GD): This is an administrative discharge based on some failure to adapt sufficiently to military routine. It is often an indulgent way to remove a person without severely affecting the rest of his life in an adverse manner. It is like being released from active duty with a charitable B-.

Other Than Honorable Discharge (OTHD): This administrative discharge is the severest that can be issued for non-court-martial offenses. These are usually processed for unacceptable behavior and violations such as security issues, civilian law charges, drug use, assaults, etc. It is like being released from active duty with a C-.

Bad Conduct Discharge (BCD): This discharge (for enlisted only) is usually the result of a court-martial conviction and depending on severity of the crime it may include prison time. It is like being released from active duty with a D-.

Dishonorable Discharge (DD): For enlisted only. The most punitive of discharges for felony crimes such as murder or desertion or espionage. This can be followed by prison time or execution. It is like being released from active duty with an F.

OTHER TYPES OF SEPARATIONS FROM MILITARY SERVICE

Entry-Level Separation: This type of estrangement is given to a recruit or basic trainee who is incapable of satisfactorily completing training. Since these are convened within 180 days of initial service, the individual is not considered in permanent military status, and therefore not entitled to any of the benefits bestowed upon veterans.

Medical Separation: This separation may be awarded military personnel who are injured or become sick to such a degree that they can no longer perform their assigned duties.

For the Convenience of the Government Separation: This is a rare, and murky, class of separation from military service, sometimes due to budget cutbacks. Each branch of the armed forces has their own protocol.

NOTE: Many seriously wounded-in-combat individuals are medically-retired. This is not a Discharge. It is a retirement. Retired persons do not receive a separation nor a discharge.

UNIFORM CODE OF MILITARY JUSTICE (UCMJ)

The UCMJ encompasses all things law in the military. It provides a standardized system of rules and regulations, and penalties for the violations of its edicts. The Code is massive, but highlighted below are just two areas for summation as these were the most common to military personnel.

CRIME & PUNISHMENT

Non-judicial Punishment (NJP): This type of penalty is meted out under Article 15 of the UCMJ. The hearing is conducted by the person's own Commanding Officer and includes such minor crimes as reporting late, petty theft, lying, sleeping on (non-combat) guard duty, etc. Punishments dispensed can be lenient to severe. A recipient of the punishment may be confined with limited rations, be assigned extra duties, receive a monetary fine, be reduced in rank, or let off with just a reprimand. NJP is often referred to as Office Hours or Captain's Mast.

Court Martial: This type of judicial process is far more serious, with major consequences. There are three types:

1) **Summary Court-Martial:** This type of trial is only for enlisted personnel who have violated more serious crimes than Article 15 infractions, or have repeated less serious crimes, or have refused NJP. It is a trial by one Commissioned Officer serving as judge and jury. It is similar to a civilian trial except that the defendant does not have the right to a free attorney but can hire one on their own. Penalties vary by pay grade, but can be up to of one-month confinement, hard labor, loss of pay, and/or loss of rank.

2) **Special Court-Martial:** This is a misdemeanor court and it tries both enlisted and officers. Defendants choose to be heard by a military judge alone or they can request a panel of three or more persons, plus a military judge. Enlisted accused may request ⅓ of the panel be enlisted personnel. The accused is entitled to a free military attorney and can hire a civilian attorney on his own. Again, penalties very by pay grade, but also depend on whether the guilty party is enlisted or an officer. Punishments can be up to one year of confinement, hard labor, or loss of 2/3 pay for up to a year. Additionally, for enlisted personnel only, demotion to Private and a Bad Conduct Discharge are possible. Officers found guilty cannot lose rank nor be discharged in this court.

3) **General Court-Martial:** This is the most serious of courts and reserved for felonies. This highest of military courts tries both enlisted and officers. Unless a defendant waives it, a pre-trial, much like a civilian grand jury process, must take place before a General Court-Martial trial may begin. Defendants choose to be heard by a military judge alone or they can request a panel of five or more persons, plus a military judge. Enlisted accused may request ⅓ of the panel be enlisted personnel. The accused is entitled to a free military attorney and can hire a civilian attorney on his own. Penalties are often severe. Almost any punishment may be meted out including Dishonorable Discharge (for enlisted), dismissal (for officers), and even the death penalty.

MEMORANDA

MEMORANDA

MEMORANDA